Student's Solutions Manual

Triola

Elementary Statistics

Sixth Edition

Student's Solutions Manual

Triola

Elementary Statistics

Sixth Edition

Milton Loyer
Penn State University

Addison-Wesley Publishing Company

Reading, Massachusetts • Menlo Park, California • New York
Don Mills, Ontario • Wokingham, England • Amsterdam • Bonn
Sydney • Singapore • Tokyo • Madrid • San Juan • Milan • Paris

Reprinted with corrections, February 1995.

Copyright © 1995 by Addison-Wesley Publishing Company, Inc.

ISBN 0-201-57683-X
     5 6 7 8 9 10-BAH-9796

PREFACE

This manual contains the solutions to the odd-numbered problems of the textbook Elementary Statistics, Sixth Edition, by Mario Triola. In the worked problems, intermediate steps are provided for the calculations. When appropriate, additional hints and comments are included and prefaced by NOTE.

Many statistical problems are best solved using particular formats. Recognizing and following these patterns promotes understanding and develops the capacity to apply the concepts to other problems. This manual identifies and employs such formats whenever practicable.

For best results, read the text carefully before attempting the exercises and attempt the exercises before consulting the solutions. This manual has been prepared to provide a check and extra insights for exercises that have already been completed and to provide guidance for solving exercises that have already been attempted but have not been successfully completed.

I would like to thank Mario Triola for writing an excellent elementary statistics book and for inviting me to prepare this solutions manual.

TABLE OF CONTENTS

Chapter 1: Introduction to Statistics...1

Chapter 2: Descriptive Statistics.....................................7

Chapter 3: Probability..38

Chapter 4: Probability Distributions...........................54

Chapter 5: Normal Probability Distributions.......................70

Chapter 6: Estimates and Sample Sizes.........................92

Chapter 7: Hypothesis Testing................................102

Chapter 8: Inferences from Two Samples.........................125

Chapter 9: Correlation and Regression.........................150

Chapter 10: Multinomial Experiments and Contingency Tables.........173

Chapter 11: Analysis of Variance..............................185

Chapter 12: Statistical Process Control.........................204

Chapter 13: Nonparametric Statistics...........................215

Chapter 1

Introduction to Statistics

1-2 The Nature of Data

1. discrete, since the number of defectives must be an integer

3. continuous, since weight can be any value on a continuum

5. continuous, since speed can be any value on a continuum

7. discrete, since the number recognizing the name must be an integer

9. ordinal, since the labels give relative position in a hierarchy

11. nominal, since the color is used only to distinguish the pieces

13. nominal, since the numbers are used for identification only. Even though they are assigned alphabetically within regions, zip codes are merely numerical names for post offices.
    NOTE: Suppose the numbers had resulted from placing all the post offices in one large list alphabetically (or by city size, or by mail volume, etc.), however, so that 17356 [Red Lion, PA] was the 17,356th post office in the list; then zip codes, like the order of finishers in a race, would illustrate the ordinal level of measurement.

15. ratio, since differences are meaningful, and there is an inherent zero

17. interval, since differences are meaningful but ratios are not

19. a. If a person with an IQ of 150 is twice as intelligent as a person with an IQ of 75, then IQ is being measured on a ratio scale.
    b. If a person with an IQ of 150 is not twice as intelligent as a person with an IQ of 75, then IQ is not being measured on a ratio scale. If the difference in intelligence between persons with IQ's of 75 and 100 is the same as the difference between persons with IQ's of 150 and 175, then IQ is being measured on an interval scale.

21. Years are not data at the ratio level of measurement because the year zero has been arbitrarily assigned so that 1900 does not represent twice as much time as 950 -- consider, for example, the Chinese numerical representations for the given years. Since the time difference between 1900 and 1920 is the same as the time difference between 1920 and 1940, however, years are data at the interval level of measurement.

1-3 Uses and Abuses of Statistics

1. Studies sponsored by groups with a desire to show a particular result are sometimes biased toward that result. Such bias is not necessarily intentional and may owe to well-intentioned efforts to control the experiment by eliminating subjects and/or situations not conducive to the desired result, or to a subconscious tendency to interpret unclear data in the hoped-for direction. In addition, studies that do not support the desired conclusion will probably be re-done and/or not be reported.

3. Because the graph does not start at zero, the given sizes are not in the same ratio as the numbers they represent and give an exaggerated impression of the differences.

5. When there appears to be a cause-and-effect relationship between two factors, that relationship may be created by an unobserved third factor; or the cause-and-effect may even work in the reverse direction. There may, for example, be a genetic factor that tends to produce individuals that are both lower-scoring and inclined to physically addictive behaviors; or it may be that getting lower grades causes students to become more nervous and more inclined toward physically addictive behaviors.

7. The study showed that Corvette drivers have a higher death rate than Volvo drivers, not necessarily that Corvettes are inherently more deadly and unsafe than Volvos. It could be, for example, that people with reckless personalities tend to purchase Corvettes and that they would have just as many accidents no matter what type of car they drove.

9. Alumni with low (or no) annual salaries would probably feel discouraged and be less inclined to respond than would those with high annual salaries. This means that the responses would not represent all alumni and would tend to overstate the true typical annual salary. Other factors that might affect the results include (1) alumni wanting to make themselves look good might overstate annual salary, (2) alumni with income from unreputable sources and/or who understated income on tax forms might understate annual salary, (3) alumni wishing to avoid pleas for donations might understate annual incomes.

11. a. The pay cut was 20% of $400, or $80, leaving a weekly salary of $320.
    b. The pay increase was 20% of $320, or $64, producing a weekly salary of $384.
    c. The percentages were calculated on the various current salaries and did not produce cuts and raises to return the employee to her original salary.

13. Statement (a) claims a specific cause-and-effect relationship, while the more general statement (b) merely states that the factors are related. Because statement (a) goes beyond the facts to reach a conclusion, it is open to challenge. It could be argued, for example, that people under the influence of alcohol don't actually cause crashes but might be too impaired to avoid involvement in accidents caused by others.

15. The purpose of a graph is to convey information visually. Since there is no correspondence between the lengths of the bars and the amounts they represent (the 983.5 bar, for example, is over twice the size of the 643.3 bar), the bars are misleading and actually convey misinformation that hinders rather than helps the reader to receive the proper values.

17. Assuming that each of the 20 individual subjects is ultimately counted as a success or not (i.e., that there are no "dropouts" or "partial successes"), the success rates in fraction form must be one of 0/20, 1/20, 2/20,..., 19/20, 20/20. In percentages, these rates are multiples of 5 (0%, 5%, 10%,..., 95%, 100%), and values such as 53% and 58% are not mathematical possibilities.

19. a. Since 100% is the totality of whatever is being measured, removing 100% of some quantity means that none of it is left.

   b. Reducing plaque by over 300% would mean removing three times as much plaque as is there, and then removing even more!

1-4 Methods of Sampling

1. systematic, since pagers were selected at regular intervals

3. cluster, since the entire set of interest (assumed to be all Ohio State University students) was divided into sections, and all the members in the selected sections were surveyed. NOTE: An added complication is the fact that the clusters overlap; any student in more than one class could conceivably be in two different selected clusters and be counted twice -- whether or not that creates a problem depends on the nature of the survey and may require advanced techniques. Notice also the stated assumption; if the set of interest is all U.S. college students, for example, the dean is using convenience sampling (because he is using a readily available group -- his own university), with cluster sub-sampling to reduce the sample to a manageable size.

5. convenience, since the sample was simply those choosing to respond

7. stratified, since the set of interest was divided into subpopulations from which the actual sampling was done

9. cluster, since the entire set of interest (all leukemia patients) was divided into counties, and all the appropriate persons in each selected county were interviewed

11. random, since each company has an equal chance of being selected

13. People probably don't know their waist size precisely and would respond with estimates and/or rounded values. In addition, people would probably tend to give themselves the benefit of the doubt and respond with values smaller than the true measurement.

15. a. Open questions elicit the respondent's true feelings without putting words or ideas into his mind. In addition, open questions might produce responses the pollster failed to consider. Unfortunately open questions sometimes produce responses that are rambling, unintelligible or not relevant.

   b. Closed questions help to focus the respondent and prevent misinterpretation of the question. Sometimes, however, closed questions reflect only the wording and opinions of the pollster and do not allow respondents to express legitimate alternatives.

   c. Closed questions are easier to analyze because the pollster can control the number of possible responses to each question and word the responses to establish relationships between questions.

1.5 Statistics and Computers

1. ```
MTB > SET C1
DATA> 2 4 1 2 3 2 3 1
DATA> END
MTB > PRINT C1
```

   ```
   C1
      2   4   1   2   3   2   3   1
   ```

2. [Even though it bears an even number, this problem is included because it is used in subsequent odd problems. In general, such will be the procedure throughout this manual.]
   ```
   MTB > SET C2
   DATA> 3.2 22.6 23.1 16.9 0.4 6.6 12.5 22.8
   DATA> END
   MTB > PRINT C2
   ```

   ```
   C2
      3.2   22.6   23.1   16.9   0.4   6.6   12.5   22.8
   ```

3. a. ```
   MTB > LET C3 = C1 + C2
   MTB > PRINT C3
   ```

   ```
   C3
      5.2   26.6   24.1   18.9   3.4   8.6   15.5   23.8
   ```

   Each value in C3 (column 3) is the sum of the corresponding values in C1 and C2. NOTE:
   Even though each set of values is designated as a column, Minitab prints the values of a
   single column in row format in order to save space. As the next exercises indicate, Minitab
   uses the column format when printing more than one column. Since C1 and C2 each contain
   8 values, C3 contains 8 sums; when C1 and C2 do not contain the same number of values,
   Minitab finds as many sums as possible and provides an appropriate message.
   b. No, the values in C3 are of no practical significance. Calculators and computers give
   numerical values without considering whether they make any sense; it is the always
   the user's responsibility to consider the operations being performed and the units involved
   (e.g., feet, pounds, mpg, etc.).

4. ```
READ> C4 C5 C6
DATA> 623 509 2.6
DATA> 454 471 2.3
DATA> 643 700 2.4
DATA> 585 719 3.0
DATA> 719 710 3.1
DATA> END
    5 ROWS READ
MTB > PRINT C4 C5 C6
```

   ```
   ROW   C4   C5   C6

    1   623   509   2.6
    2   454   471   2.3
    3   643   700   2.4
    4   585   719   3.0
    5   719   710   3.1
   ```
   ```

5. MTB > SAVE 'EXER'
   MTB > STOP
   [The computer will print a statement about ending Minitab.]

   MTB > RETRIEVE 'EXER'
   [The computer will print details about retrieving the file.]
   MTB > PRINT C1-C6

ROW	C1	C2	C3	C4	C5	C6
1	2	3.2	5.2	623	509	2.6
2	4	22.6	26.6	454	471	2.3
3	1	23.1	24.1	643	700	2.4
4	2	16.9	18.9	585	719	3.0
5	3	0.4	3.4	719	710	3.1
6	2	6.6	8.6			
7	3	12.5	15.5			
8	1	22.8	23.8			

NOTE: In general, Minitab is very user-friendly and follows normal human conventions. Within a PRINT command, Minitab assumes the hyphen means "through"; within a LET command, Minitab assumes the hyphen means "minus."

7. MTB > SET C7
   DATA> 25(12.345)
   DATA> END
   MTB > PRINT C7

C7
```
12.345   12.345   12.345   12.345   12.345   12.345   12.345
12.345   12.345   12.345   12.345   12.345   12.345   12.345
12.345   12.345   12.345   12.345   12.345   12.345   12.345
12.345   12.345   12.345   12.345
```

Minitab prints the 25 12.345 values located in column 7.

9. MTB > SET C11
   DATA> 2 4 1 2 3 2 3 1
   DATA> END
   a. MTB > MEAN C11
      MEAN   =   2.2500
   b. MTB > MAXIMUM C11
      MAXIMUM =   4.0000
   c. MTB > MINIMUM C11
      MINIMUM =   1.0000
   d. MTB > SUM C11
      SUM   =   18.000
   e. MTB > SSQ C11
      SSQ   =   48.000

The commands in parts (a)-(e) above produce the arithmetic average of the values, the largest value, the smallest value, the sum of the values, and the sum of the squares of the values. NOTE: Statisticians use the word "mean" for "arithmetic average." This is actually a mathematical version of an historic use of the word, for in very early English "mean" was defined as "average" in the sense of "commonplace" or "not superior." The Christmas song *What Child Is This?* asks, for example, "Why lies he in such mean estate, where ox and ass are feeding?" The SSQ stands for $\underline{S}$um of $\underline{SQ}$uares and is $2^2 + 4^2 + \ldots 1^2 = 48$.

Review Exercises

1. a. ordinal, the categories give relative rankings but differences between categories are not necessarily consistent
   b. ratio, differences between values are consistent and there is a meaningful zero
   c. ordinal, the categories give relative rankings but differences between categories are not necessarily consistent
   d. nominal, the categories give names only and not any ordering
   e. interval, differences between values are consistent but there is no meaningful zero

3. The police had to estimate the quantity of albums seized; assuming they did not count each album, this could have been done by volume, weight, number of boxes, etc. They also had to estimate the value per album; this could have been done using retail price, "street" price, wholesale price, actual cost to produce, etc. They might tend to exaggerate the value of the albums and/or use the method that produces the largest value in order to appear more effective or in hopes that the magnitude of the criminal activity might justify an increase in manpower, salaries, etc.

5. a. systematic, since products are selected at regular intervals
   b. random, since each car has the same chance of being selected
   c. cluster, since the stocked items were organized in stores and all the items in randomly selected stores were chosen
   d. stratified, since drivers were classified by sex and age in order to make the selections
   e. convenience, since the sample was composed of those who chose to take a test drive.
   NOTE: If the car maker first selected the particular dealership at random from among all its dealerships, then cluster sampling was used. Since the apparent set of interest is all potential customers and not just all who take a test drive, the selections at the local dealership were made by convenience sampling.

7. When giving their own personal data (height, weight, income, age, etc.) people sometimes give round numbers as a subconscious way to avoid invasions of privacy (and they often round in the most favorable direction). When answering for someone else, people give round numbers because they may not know the exact value.

Chapter 2

Descriptive Statistics

2-2 Summarizing Data

1. Subtracting two consecutive lower class limits indicates that the class width is 88 - 80 = 8. Since there is a gap of 1.0 between the upper class limit of one class and the lower class limit of the next, class boundaries are determined by increasing or decreasing the appropriate class limits by (1.0)/2 = 0.5. The class boundaries and class marks are given in the following table.

IQ	class boundaries	class mark	frequency
80- 87	79.5 - 87.5	83.5	16
88- 95	87.5 - 95.5	91.5	37
96-103	95.5 - 103.5	99.5	50
104-111	103.5 - 111.5	107.5	29
112-119	111.5 - 119.5	115.5	14
			146

NOTE: Although they often contain extra decimal points and may involve consideration of how the data were obtained, class boundaries are the key to tabular and pictorial data summaries. Once the class boundaries are obtained, everything else falls into place. In this case, the first class width is readily seen to be 87.5 - 79.5 = 8.0 and the first class mark is (79.5 + 87.5)/2 = 83.5. In this manual, class boundaries will typically be calculated first and then be used to determine other values. In addition, since the sum of the frequencies (i.e., the total number of values) is an informative number and used in many subsequent calculations, it will typically be shown as an integral part of each table.

3. Since the gap between classes is 0.1, the appropriate class limits are increased or decreased by (0.1)/2 = .05 to obtain the class boundaries and the following table.

Weight (kg)	class boundaries	class mark	frequency
16.2-21.1	16.15 - 21.15	18.65	16
21.2-26.1	21.15 - 26.15	23.65	15
26.2-31.1	26.15 - 31.15	28.65	12
31.2-36.1	31.15 - 36.15	33.65	8
36.2-41.1	36.15 - 41.15	38.65	3
			54

The class width is 21.15 - 16.15 = 5.00 and the first class mark is (16.15 + 21.15)/2 = 18.65.

5. The relative frequency for each class is found by dividing its frequency by 146, the sum of the frequencies. NOTE: As before, the sum is included as an integral part of the table. For relative frequencies, this should always be 1.000 (i.e., 100%) and serves as a check for the calculations.

IQ	relative frequency
80- 87	.110
88- 95	.253
96-103	.342
104-111	.199
112-119	.096
	1.000

7. The relative frequency for each class is found by dividing its frequency by 54, the sum of the frequencies. NOTE: In #5, the relative frequencies were expressed as decimals; here they are expressed as percents The choice is arbitrary.

Weight (kg)	relative frequency
16.2-21.1	29.6%
21.2-26.1	27.8%
26.2-31.1	22.2%
31.2-36.1	14.8%
36.2-41.1	5.6%
	100.0%

9. The cumulative frequencies are determined by repeated addition of successive frequencies to obtain the combined number in each class and all previous classes. NOTE: Consistent with the emphasis that has been placed on class boundaries, we choose to use upper class boundaries in the "less than" column. It is assumed that intelligence occurs on a continuum and that IQ points are the nearest whole number representation of a person's true measure of intelligence. An IQ of 87.9, for example, would be reported as 88 and fall in the second class. The 16 IQ's in the first class, therefore, are better described as being "less than 87.5" (using the upper class boundary) than as being "less than 88." This distinction becomes crucial in the construction of pictorial representations in the next section. In addition, the fact that the final cumulative frequency must equal the total number (i.e, the sum of the frequency column) serves as a check for calculations. The sum of cumulative frequencies, however, has absolutely no meaning and is not included.

IQ	cumulative frequency
less than 87.5	16
less than 95.5	53
less than 103.5	103
less than 111.5	132
less than 119.5	146

11. The cumulative frequencies are determined by repeated addition of successive frequencies to obtain the combined number in each class and all previous classes.

Weight (kg)	cumulative frequency
less than 21.15	16
less than 26.15	31
less than 31.15	43
less than 36.15	51
less than 41.15	54

13. There is more than one acceptable solution. One possibility is to note that for a range of .95 - 0.26 = 4.69 to be covered with 10 classes, there must be at least (4.69)/10 = .469 units per class. Rounding up to a class width of .50 and starting at .25, for example, produces a first class with lower and upper class limits of .25 and .74. NOTE: The second class would then have lower and upper class limits of .75 and 1.24.

15. There is more than one acceptable solution. One possibility is to note that for a range of 4367 - 1650 = 2717 to be covered with 7 classes, there must be at least (2717)/7 = 388 units per class. Rounding up to a class width of 400 and starting at 1600, for example, produces a first class with lower and upper limits of 1600 and 1999. NOTE: The second class would then have lower and upper limits of 2000 and 2399.

17. The class limits determined in exercise 13 produce the table given at the right.

Weight (lbs)	frequency
0.25 - 0.74	2
0.75 - 1.24	8
1.25 - 1.74	17
1.75 - 2.24	8
2.25 - 2.74	8
2.75 - 3.24	7
3.25 - 3.74	7
3.75 - 4.24	1
4.25 - 4.74	3
4.75 - 5.24	7
	62

19. The class limits determined in exercise 15 produce the table given at the right.

Weight (lbs)	frequency
1600 - 1999	1
2000 - 2399	1
2400 - 2799	12
2800 - 3199	6
3200 - 3599	7
3600 - 3999	4
4000 - 4399	1
	32

NOTE: Starting with 1600 instead of the lowest score of 1650 makes the lower class limits in 100's instead of 50's and makes the data more "centered" --i.e., the lowest and highest values of 1650 and 4367 are 50 and 32 units in from the limits of the table.

21. The total numbers (i.e., sums of the frequency columns) for the men and women were 13,055 and 721 respectively. The relative frequency tables are as follows.

ethanol consumed by men (oz.)	relative frequency
0.0 - 0.9	.019
1.0 - 1.9	.071
2.0 - 2.9	.118
3.0 - 3.9	.171
4.0 - 4.9	.087
5.0 - 9.9	.273
10.0 - 14.9	.142
15.0 or more	.118
	1.000

ethanol consumed by women (oz.)	relative frequency
0.0 - 0.9	.010
1.0 - 1.9	.072
2.0 - 2.9	.173
3.0 - 3.9	.265
4.0 - 4.9	.042
5.0 - 9.9	.279
10.0 - 14.9	.060
15.9 or more	.100
	1.000

The distributions seem very similar except that there are proportionately more women in the 2.0-3.9 range and proportionately less women in the 10.0-14.9 range. NOTE: Due to rounding, the relative frequencies given actually sum to 0.999 for the men and to 1.001 for the women. Discrepancies of 1 or 2 at the last decimal place are probably due to rounding; larger discrepancies should not be so attributed, and such work should be carefully checked.

23. a. classes mutually exclusive? yes
    b. all classes included? yes
    c. same width for all classes? no
    d. convenient class limits? yes
    e. between 5 and 20 classes? no
    Two of the five guidelines were not followed. NOTE: This does not mean the table is in error. The guidelines are only suggestions that make most presentations more readable; depending on the context, the given table may be the best way to present the ages.

2-3 Pictures of Data

1. See the figure below. The bars extend from class boundary to class boundary. Each axis is labeled numerically <u>and</u> with the name of the quantity represented.

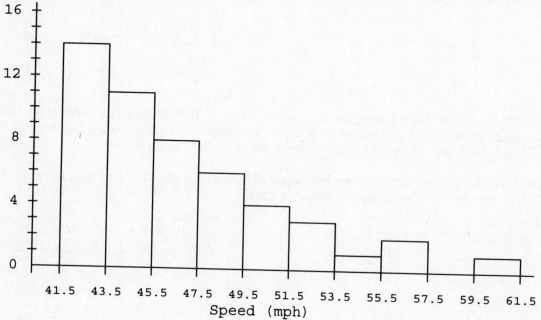

3. See the figure below. The bars extend from class boundary to class boundary. The relative frequencies are the original frequencies divided by 40, the sum of the frequencies. Boundaries were determined assuming the reported ages were rounded to the nearest tenth of a year.

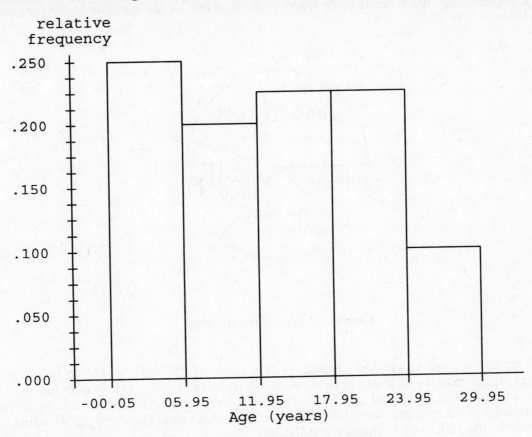

5. See the figure below, with bars arranged in order of magnitude.

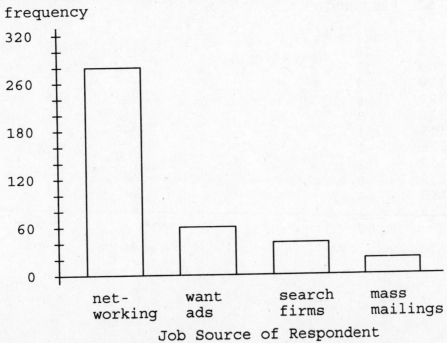

7. See the figure below. The sum of the frequencies is 50; the relative frequencies are 23/50 = 46%, 9/50 = 18%, 12/50 = 24%, and 6/50 = 12%. The corresponding central angles are (.46)360° = 165.6°, (.18)360° = 64.8°, (.24)360° = 86.4°, and (.12)360° = 43.2°. NOTE: To be complete, the figure needs to be titled with the name of the quantity being measured.

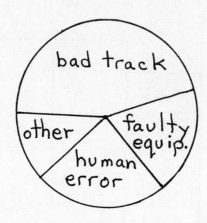

Causes of Train Derailments

9. There is more than one acceptable solution. For a range of 193 - 72 = 121 to be covered with 12 classes, there must be at least (121)/12 = 10.08 units per class. One possibility is to round <u>down</u> to a class width of 10 and use 70 as the first lower class limit. While this produces 13 classes instead of 12, the added readability created by such round numbers justifies that choice and results in the following frequency distribution.

length (minutes)	frequency
70 - 79	1
80 - 89	2
90 - 99	18
100 - 109	17
110 - 119	7
120 - 129	6
130 - 139	3
140 - 149	1
150 - 159	2
160 - 169	2
170 - 179	0
180 - 189	0
190 - 199	1
	60

a. See the figure below. The bars extend from class boundary to class boundary.

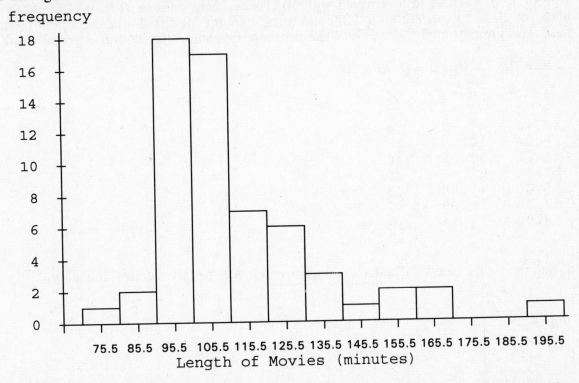

frequency

Length of Movies (minutes)

NOTE: The bars always extend from class boundary to class boundary, but the numerical labels along the horizontal axis may be placed at the class midpoints (as above) or at any other convenient points.

b. The figure shows that the data are positively skewed.

11. There is more than one acceptable solution. One possibility is to note that for a range of
20.58 - 1.65 = 18.93 to be covered with 10 classes, there must be at least (18.93)/10 = 1.893
units per class. Rounding up to 2.00 and using 1.00 for the first lower class limit produces
first class limits from 1.00 to 2.99. The resulting frequency distribution is given below.

weight (lbs)	frequency
1.00 - 2.99	4
3.00 - 4.99	4
5.00 - 6.99	12
7.00 - 8.99	9
9.00 - 10.99	12
11.00 - 12.99	10
13.00 - 14.99	5
15.00 - 16.99	3
17.00 - 18.99	1
19.00 - 20.99	2
	62

a. See the figure below. The bars extend from class boundary to class boundary.

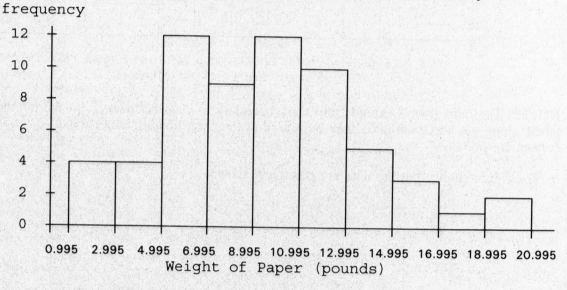

b. The figure shows that the data have a slight positive skew.

13. There is more than one acceptable solution. One possibility is to group the numbers naturally by fives to produce the frequency distribution given below. NOTE: Since zero was not one of the possible numbers, the first class contains only 4 achievable values; each of the other classes contains 5 achievable values.

number	frequency
0 - 4	23
5 - 9	34
10 - 14	26
15 - 19	30
20 - 24	33
25 - 29	31
30 - 34	30
35 - 39	34
40 - 44	31
45 - 49	28
	300

a. See the figure below. The bars extend from class boundary to class boundary.

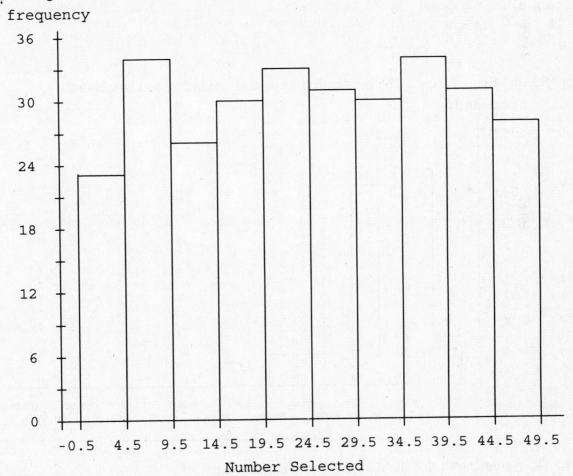

b. The figure shows that the data are approximately uniform.

15. There is more than one acceptable solution. One possibility is to note that for a range of 157.8 - 3.6 = 154.2 to be covered with 16 classes, there must be at least (154.2)/16 = 9.64 units per class. Rounding up to a class width of 10 and using 0 for the first lower class limit produces the frequency distribution given below.

income		frequency
0.0 -	9.9	6
10.0 -	19.9	27
20.0 -	29.9	31
30.0 -	39.9	25
40.0 -	49.9	21
50.0 -	59.9	4
60.0 -	69.0	4
70.0 -	79.9	2
80.0 -	89.9	3
90.0 -	99.9	0
100.0 -	109.9	1
110.0 -	119.9	0
120.0 -	129.9	0
130.9 -	139.9	0
140.0 -	149.9	0
150.0 -	159.9	1
		125

a. See the figure below. The bars extend from class boundary to class boundary.

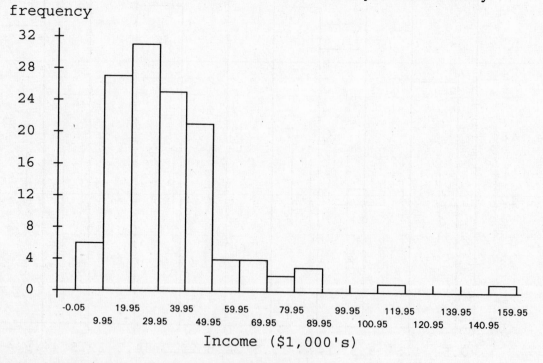

b. The figure shows that the data are positively skewed.

17. See the figure below. NOTE: The graph returns to a frequency of zero at the midpoints of the classes before and after the given data.

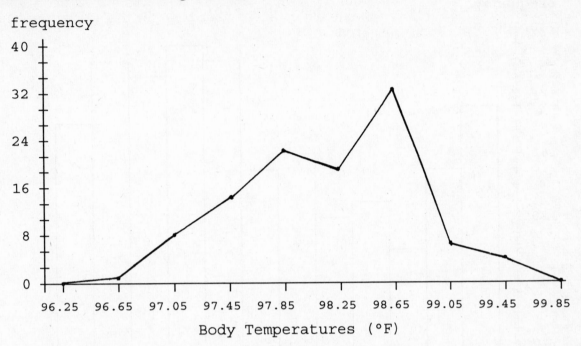

19. a. The two figures are given below. The histogram for π represents a more even distribution.

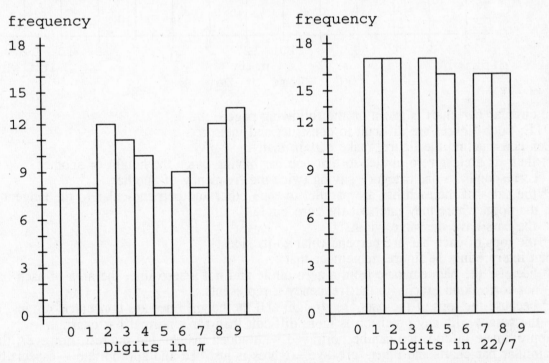

b. The number π is irrational (i.e., it cannot be represented as the quotient of two whole numbers). The decimal representation of the rational number 22/7 is obtained by dividing 7 into 22; because there are a maximum of 6 possible remainder digits at each step (1,2,3,4,5,6) that keep generating digits in the quotient, there are a maximum of six possible quotient digits that will repeat for ever in a regular cycle.

21. a. Here is the bar chart.

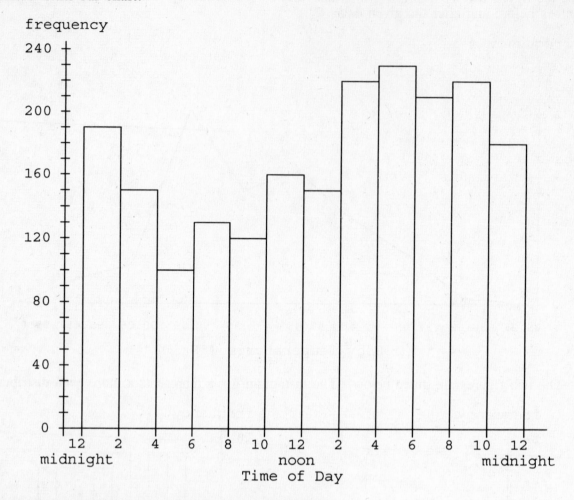

The circular bar chart is given on the following page.
NOTE: Such figures are difficult to construct and interpret.
When constructing the figure, make certain that...
  * there is a clear zero circle, so that one bar having twice the height of another
    corresponds to one category having twice the frequency of another.
  * the sides of the each bar are parallel to each other and perpendicular to the tangent line at
    the point where they intersect the zero circle.
  * the bars have the same width.
  * the tops of each bar are perpendicular to its sides.
When interpreting the figure, remember that ...
  * because the bars emanate from a zero circle and not a zero line, the area of each bar will
    not correspond exactly to the frequency it represents.
  * because the tops of the bars are straight and the height lines are concentric circles, reading
    the height with any precision is more difficult than for a normal bar chart.
If approximate information can be portrayed with added insight and interest, however, the use
of circular bar charts and other "creative" figures is justified and appropriate -- especially in
informal settings.

Number of Traffic Fatalities

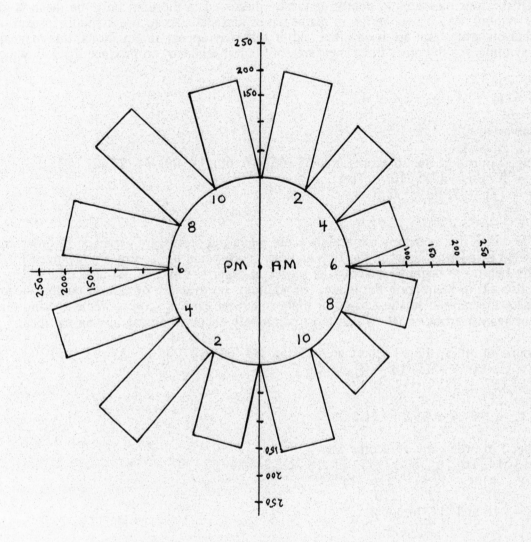

b. Because the circular bar chart has no beginning and no end, it does not artificially center on
   noon and better conveys the pattern in the numbers of death across the midnight hours. It
   might be argued, however, that the traditional bar chart is easier to read, more accurate,
   much less difficult to construct, and that adequate labeling allows the reader to appreciate the
   continuity in the time scale.

c. The fact that there are the least number of fatal crashes between 4 and 6 am does not
   necessarily make it the safest time to drive. That's like saying the best place to sit when
   you're sick is on the top of a piano because very, very few ever die while sitting on a piano.
   There are fewer cars on the road during those hours, and it that sense it may be considered a
   wise and safe time to travel. One must consider the rates of fatal accidents, however, as
   well as the numbers of such occurrences -- it could be, for example, that fatal crashes
   involve 1.0% of all the cars on the road during the 4-6 am slack period but only 0.1% of the
   many cars on the road during the 4-6 pm rush hour.

2-4 Measures of Central Tendency

NOTE: Of the four measures of central tendency discussed in the text, only the mean is given a particular symbol. As it is common in mathematics and statistics to use symbols instead of words to represent quantities that are used often and/or that may appear in equations, this manual also employs symbols for the other three measures of central tendency to produce the following notation:

mean = $\bar{x}$
median = m
mode = M
midrange = m.r.

1. Arranged in order, the 10 scores are: 65 66 67 68 71 73 74 77 77 77
   a. $\bar{x} = (\Sigma x)/n = (715)/10 = 71.5$
   b. m = (71 + 73)/2 = 72.0
   c. M = 77
   d. m.r. = (65 + 77)/2 = 71.0
   NOTE: Since the median is the middle score <u>when the scores are arranged in order</u> and the midrange is halfway between the first and last score <u>when the scores are arranged in order</u>, it is usually helpful to begin by placing the scores in order -- this will not affect the mean, and it may also aid in identifying the mode. In addition, no measure of central tendency can have a value lower than the smallest score or higher than the largest score -- remembering this helps to protect against gross errors, which most commonly occur when calculating the mean.

2. Arranged in order, the 10 scores are: 42 54 58 62 67 77 77 85 93 100
   a. $\bar{x} = (\Sigma x)/n = (715)/10 = 71.5$
   b. m = (67 + 77)/2 = 72.0
   c. M = 77
   d. m.r. = (42 + 100)/2 = 71.0

3. Arranged in order, the 15 scores are:
   .12 .13 .14 .16 .16 .16 .17 .17 .17 .18 .21 .24 .24 .27 .29
   a. $\bar{x} = (\Sigma x)/n = (2.81)/15 = .187$
   b. m = (71 + 73)/2 = .170
   c. M = .16 and .17 (bimodal)
   d. m.r. = (.12 + .29)/2 = .205

5. Arranged in order, the 27 scores are:
   14 16 17 17 18 18 19 20 21 23 23 24 25 25
   27 28 28 29 30 31 33 34 37 38 40 42 51
   a. $\bar{x} = (\Sigma x)/n = (728)/27 = 27.0$
   b. m = 25.0
   c. M = 17,18,23,25,28 (multimodal)
   d. m.r. = (14 + 51)/2 = 32.5
   NOTE: The purpose of descriptive statistics in general and measures of central tendency in particular is to provide a meaningful summary of the data, and the above determination of the mode followed the definition quite literally. From a practical point of view, most statisticians would probably say there was no meaningful mode. Grouping by ten's, one could say the mode was in the 20's -- i.e., more people died in their 20's than in any other 10-year group.

7. Arranged in order, the 8 scores are: 0.63 0.92 1.40 1.41 1.74 2.10 2.19 2.87
   a. $\bar{x}$ = (Σx)/n = (13.26)/8 = 1.6575
   b. m = (1.41 + 1.74)/2 = 1.575
   c. M = [none]
   d. m.r. = (0.63 + 2.87)/2 = 1.750
   NOTE: The mean was 1.6575; according to the rule given in the text, this value should be
   rounded to three decimal places. While the text describes how many decimal places to present
   in an answer, it does not describe the actual rounding process. When the figure to be rounded
   is exactly half-way between two values (i.e., the digit in the position to be discarded is a 5, and
   there are no further digits because the calculations have "come out even"), there is no
   universally accepted rule. Some authors say to always round up a 5; others correctly note that
   always rounding up a value exactly half-way between introduces a consistent bias -- the value
   should be rounded up half the time and rounded down half the time. In such cases, some
   authors suggest rounding toward the even value (e.g., .65 becomes .6 and .75 becomes .8)
   and other simply suggest flipping a coin. In this manual, the results of calculations producing
   values exactly half-way between will be reported without rounding (i.e., stated to one more
   decimal than usual).

9. Arranged in order, the 12 scores are:
   654.2 661.3 662.2 662.7 667.0 667.4 669.8 670.7 672.2 672.2 672.6 679.2
   a. $\bar{x}$ = (Σx)/n = (8011.5)/12 = 667.625
   b. m = (667.4 + 669.8)/2 = 668.60
   c. M = 672.2
   d. m.r. = (654.2 + 679.2)/2 = 666.70

11. Arranged in order, the 10 scores are:
    2360 2526 2784 2863 2992 3253 3315 3536 3766 3784
    a. $\bar{x}$ = (Σx)/n = (31179)/10 = 3117.9
    b. m = (2992 + 3253)/2 = 3122.5
    c. M = [none]
    d. m.r. = (2360 + 3784)/2 = 3072.0

13. Construct a stem-and-leaf plot [see section 2.7] to arrange the 38 scores in order.

96.	2 6
97.	0 2 3 4 4 4 4 5 6 7 8
98.	0 0 0 0 1 2 2 2 2 4 4 6 7 7 8 8 8 8 9 9
99.	0 0 0 2 4

    a. $\bar{x}$ = (Σx)/n = (3728.8)/38 = 98.13
    b. m = (98.2 + 98.2)/2 = 98.20
    c. M = 97.4, 98.0, 98.2, 98.8 (multi-modal)
    d. m.r. = (96.2 + 99.4)/2 = 97.80

15. Construct a stem-and-leaf plot [see section 2.7] to arrange the 62 scores in order.

```
 1. | 65
 2. | 41 61 80
 3. | 26 27 45 69
 4. |
 5. | 86 87 88
 6. | 05 16 33 38 44 67 68 96 98
 7. | 57 72 98
 8. | 08 26 72 78 82 96
 9. | 09 19 41 45 46 55 64 83 92
10. | 00 58 99
11. | 03 08 36 42
12. | 29 32 43 45 56 73
13. | 05 11 31 61
14. | 33
15. | 09
16. | 08 39
17. | 65
18. |
19. |
20. | 12 58
```

a. $\bar{x} = (\Sigma x)/n = (584.54)/62 = 9.428$
b. $m = (9.19 + 9.41)/2 = 9.300$
c. $M =$ [none]
d. m.r. $= (1.65 + 20.58)/2 = 11.115$

17.

x	f	x·f
42.5	14	595.0
44.5	11	489.5
46.5	8	372.0
48.5	6	291.0
50.5	4	202.0
52.5	3	157.5
54.5	1	54.5
56.5	2	113.0
58.5	0	0.0
60.5	1	60.5
	50	2335.0

$\bar{x} = (\Sigma x \cdot f)/n$
$= (2335.0)/50$
$= 46.7$

19.

x	f	x·f
2.95	10	29.50
8.95	8	71.60
14.95	9	134.55
20.95	9	188.55
26.95	4	107.80
	50	532.00

$\bar{x} = (\Sigma x \cdot f)/n$
$= (532.00)/50$
$= 13.30$

21. a. Arranged in order, the 5 scores are: 108,000 179,000 206,000 236,000 236,000
$\bar{x} = (\Sigma x)/n = (965,000)/5 = 193,000$
$m = 206,000$
$M = 236,000$
m.r. $= (108,000 + 236,000)/2 = 172,000$

b. Arranged in order, the 5 scores are: 128,000   199,000   226,000   256,000   256,000
$\bar{x} = (\Sigma x)/n = (1065,000)/5 = 213,000$
m = 226,000
M = 256,000
m.r. = (128,000 + 256,000)/2 = 192,000

c. In general, adding (or subtracting) a constant k from each score will add (or subtract) k from each measure of central tendency.

d. Arranged in order, the 5 scores are: 108   179   206   236   236
$\bar{x} = (\Sigma x)/n = (965)/5 = 193$
m = 206
M = 236
m.r. = (108 + 236)/2 = 172

e. In general, dividing (or multiplying) each score by a constant k will divide (or multiply) each measure of central tendency by k.

23. a. The answers to exercises 1 and 2 are identical. Since the measures of central tendency attempt to summarize the data, the logical inference would be that the two data sets are identical -- or at least very similar.

b. The measures of central tendency alone do not necessarily reflect differences in spread (the data in exercise 2 is much more spread out; all the scores in exercise 1 were in the 60's or 70's) or difference in skewness (the data in exercise 2 is more symmetric and has a mode near the center of the scores; the scores in exercise 2 "pile up" at the high end and have the highest value for a mode).

25. In general, log $\bar{x}$ does not equal $(\Sigma \log x)/n$. In words, the log of the average is not the same as the average of the logs. This is analogous to the more straightforward fact that the square of the average is not the same as the average of the squares and illustrates the principle that the order in which operations are applied can make a difference. Consider the following example, in which the log base 10 is used.

x	log x	x^2
1	0	1
10	1	100
100	2	10000
111	3	10101

$\bar{x} = (\Sigma x)/n = 111/3 = 37$

$(\Sigma \log x)/n = 3/3 = 1$

$(\Sigma x^2)/n = 10101/3 \doteq 3367$

$\log \bar{x} = \log 37 = 1.568 \qquad 1 = (\Sigma \log x)/n$
$\bar{x}^2 = 37^2 = 1369 \qquad 3367 = (\Sigma x^2)/n$

27. The geometric mean of the five values is the fifth root of their product.
$\sqrt[5]{(1.10)(1.08)(1.09)(1.12)(1.07)} = 1.092$

29. a. The class mark of the last class is (20.5 + 25.5)/2 = 23

x	f	x·f
0	5	0
3	96	288
8	57	456
13	25	325
18	11	198
23	6	138
	200	1405

$\bar{x} = (\Sigma x \cdot f)/n$
$= (1405)/200$
$= 7.0$

b. The class mark of the last class is $(20.5 + 30.5)/2 = 25.5$

x	f	x·f
0.0	5	0.0
3.0	96	288.0
8.0	57	456.0
13.0	25	325.0
18.0	11	198.0
25.5	6	153.0
	200	1420.0

$\bar{x} = (\Sigma x \cdot f)/n$
$= (1420)/200$
$= 7.1$

c. The class mark of the last class is $(20.5 + 40.5)/2 = 30.5$

x	f	x·f
0.0	5	0.0
3.0	96	288.0
8.0	57	456.0
13.0	25	325.0
18.0	11	198.0
30.5	6	183.0
	200	1450.0

$\bar{x} = (\Sigma x \cdot f)/n$
$= (1450)/200$
$= 7.25$

In this case, because it involves relatively few scores and there is a reasonable upper limit (e.g, we know no student spends 100 hours a week studying) the interpretation of the open-ended class seems to make little difference and justifies the use of such classes when they make for significantly simpler presentations.

31. Arranged in order, the original 20 scores are:
0.0 0.0 1.0 1.5 1.7 1.8 2.0 2.0 2.1 2.1
2.3 2.4 2.4 2.9 3.3 3.4 3.7 4.4 4.4 4.5
$\bar{x} = (\Sigma x)/n = (47.9)/20 = 2.395$
a. After trimming the highest and lowest 10%, the remaining 16 scores are:
1.0 1.5 1.7 1.8 2.0 2.0 2.1 2.1 2.3 2.4 2.4 2.9 3.3 3.4 3.7 4.4
$\bar{x} = (\Sigma x)/n = (39.0)/16 = 2.44$
b. After trimming the highest and lowest 20%, the remaining 12 scores are:
1.7 1.8 2.0 2.0 2.1 2.1 2.3 2.4 2.4 2.9 3.3 3.4
$\bar{x} = (\Sigma x)/n = (28.4)/12 = 2.37$

2-5 Measures of Variation

1. NOTE: Although not given in the text, the symbol R will be used for the range throughout this manual. Remember that the range is the difference between the highest and the lowest scores, and not necessarily the difference between the last and the first values as they are listed.

x	$x - \bar{x}$	$(x-\bar{x})^2$	x^2
65	-6.5	42.25	4225
66	-5.5	30.25	4356
67	-4.5	20.25	4489
68	-3.5	12.25	4624
71	-0.5	0.25	5041
73	1.5	2.25	5329
74	2.5	6.25	5476
77	5.5	30.25	5929
77	5.5	30.25	5929
77	5.5	30.25	5929
715	0	204.50	51327

$$\bar{x} = (\Sigma x)/n$$
$$= 715/10$$
$$= 71.5$$

$R = 77 - 65 = 12$

by formula 2-4, $s^2 = \Sigma(x-\bar{x})^2/(n-1)$
$$= (204.50)/(9)$$
$$= 22.7$$
$$s = 4.8$$

by formula 2-6, $s^2 = [n(\Sigma x^2) - (\Sigma x)^2]/[n(n-1)]$
$$= [10(51327) - (715)^2]/[10(9)]$$
$$= 2045/90$$
$$= 22.7$$
$$s = 4.8$$

NOTE: When using formula 2-4, constructing a table having the first three columns shown above will help to organize the calculations and make errors less likely; in addition, verify that $\Sigma(x-\bar{x}) = 0$ before proceeding -- if such is not the case, there is an error and further calculation is fruitless. For completeness, and as a check, both formulas 2-4 and 2-6 were used above. In general, only formula 2-6 will be used throughout the remainder of this manual for the following reasons:

(1) When the mean does not "come out even," formula 2-4 involves round-off error and/or many messy decimal calculations.
(2) The quantities Σx and Σx^2 needed for formula 2-6 can be found directly and conveniently on the calculator from the original data without having to construct a table like the one above.

3. preliminary values: n = 15, $\Sigma x = 2.81$, $\Sigma x^2 = .5631$
$R = .29 - .12 = .17$
$s^2 = [n(\Sigma x^2) - (\Sigma x)^2]/[n(n-1)]$
$$= [15(.5631) - (2.81)^2]/[15(14)]$$
$$= (.5504)/210$$
$$= .003$$
$s = .051$
NOTE: The quantity $[n(\Sigma x^2) - (\Sigma x)^2]$ cannot be less than zero. A negative value indicates that there is an error and that further calculation is fruitless. In addition, find the value for s by taking the square root of the precise value of s^2 showing on the calculator display before it is rounded to one more decimal place than the original data.

5. preliminary values: $n = 27$, $\Sigma x = 728$, $\Sigma x^2 = 21{,}786$
$R = 51 - 14 = 37$
$s^2 = [n(\Sigma x^2) - (\Sigma x)^2]/[n(n-1)]$
   $= [27(21{,}786) - (728)^2]/[27(26)]$
   $= 58238/702$
   $= 83.0$
$s = 9.1$

7. preliminary values: $n = 8$, $\Sigma x = 13.26$, $\Sigma x^2 = 25.6620$
$R = 2.87 - 0.63 = 2.24$
$s^2 = [n(\Sigma x^2) - (\Sigma x)^2]/[n(n-1)]$
   $= [8(25.6620) - (13.26)^2]/[8(7)]$
   $= (29.4684)/56$
   $= .526$
$s = .725$

9. preliminary values: $n = 12$, $\Sigma x = 8011.5$, $\Sigma x^2 = 5{,}349{,}166.83$
$R = 679.2 - 654.2 = 25.0$
$s^2 = [n(\Sigma x^2) - (\Sigma x)^2]/[n(n-1)]$
   $= [12(5{,}349{,}166.83) - (8011.5)^2]/[12(11)]$
   $= (5869.71)/132$
   $= 44.47$
$s = 6.67$

11. preliminary values: $n = 10$, $\Sigma x = 31179$, $\Sigma x^2 = 99{,}425{,}707$
$R = 3784 - 2360 = 1424$
$s^2 = [n(\Sigma x^2) - (\Sigma x)^2]/[n(n-1)]$
   $= [10(99{,}425{,}707) - (31179)^2]/[10(9)]$
   $= 22127029/90$
   $= 245855.9$
$s = 495.8$

13. Refer also to exercise 13 of section 2-4.
preliminary values: $n = 38$, $\Sigma x = 3728.8$, $\Sigma x^2 = 365914.52$
a. $R/4 = (99.4 - 96.2)/4 = .80$
b. $s^2 = [n(\Sigma x^2) - (\Sigma x)^2]/[n(n-1)]$
   $= [38(365914.52) - (3728.8)^2]/[38(37)]$
   $= (802.32)/1406$
   $= .57$
   $s = .76$

15. Refer also to exercise 15 of section 2-4.
preliminary values: $n = 62$, $\Sigma x = 584.54$, $\Sigma x^2 = 6570.8216$
a. $R/4 = (20.58 - 1.65)/4 = 4.7325$
b. $s^2 = [n(\Sigma x^2) - (\Sigma x)^2]/[n(n-1)]$
   $= [62(6570.8216) - (584.54)^2]/[62(61)]$
   $= (65703.9276)/3782$
   $= 17.373$
   $s = 4.168$

17.

x	f	f·x	f·x^2
42.5	14	595.0	25287.50
44.5	11	489.5	21782.75
46.5	8	372.0	17298.00
48.5	6	291.0	14113.50
50.5	4	202.0	10201.00
52.5	3	157.5	8268.75
54.5	1	54.5	2970.25
56.5	2	113.0	6384.50
58.5	0	0.0	0.00
60.5	1	60.5	3660.25
	50	2334.0	109966.50

$$s^2 = [n(\Sigma f \cdot x^2) - (\Sigma f \cdot x)^2]/[n(n-1)]$$
$$= [50(109966.50) - (2335.0)^2]/[50(49)]$$
$$= (46100.00)/2450$$
$$= 18.8$$
$$s = 4.3$$

19.

x	f	f·x	f·x^2
2.95	10	29.50	87.0250
8.95	8	71.60	640.8200
14.95	9	134.55	2011.5225
20.95	9	188.55	3950.1225
26.95	4	107.80	2905.2100
	40	532.00	9594.7000

$$s^2 = [n(\Sigma f \cdot x^2) - (\Sigma f \cdot x)^2]/[n(n-1)]$$
$$= [40(9594.7000) - (532.00)^2]/[40(39)]$$
$$= (100764.0000)/1560$$
$$= 64.59$$
$$s = 8.04$$

21. NOTE: Because it allows for better appreciation of the concepts involved, formula 2-4 (and the suggested table) is employed for the next two exercises.

a.

x	x $-\overline{x}$	$(x-\overline{x})^2$
80	-6.5	42.25
81	-5.5	30.25
82	-4.5	20.25
83	-3.5	12.25
86	-0.5	0.25
88	1.5	2.25
89	2.5	6.25
92	5.5	30.25
92	5.5	30.25
92	5.5	30.25
865	0	204.50

$$\overline{x} = (\Sigma x)/n$$
$$= 865/10$$
$$= 86.5$$

$$R = 92 - 80 = 12$$
by formula 2-4, $s^2 = \Sigma(x-\overline{x})^2/(n-1)$
$$= (204.50)/(9)$$
$$= 22.7$$
$$s = 4.8$$

Adding a constant to each score does not affect the spread of the scores or the measures of dispersion.

b.

x	x $-\bar{x}$	$(x-\bar{x})^2$
60	-6.5	42.25
61	-5.5	30.25
62	-4.5	20.25
63	-3.5	12.25
66	-0.5	0.25
68	1.5	2.25
69	2.5	6.25
72	5.5	30.25
72	5.5	30.25
72	5.5	30.25
665	0	204.50

$$\bar{x} = (\Sigma x)/n$$
$$= 665/10$$
$$= 66.5$$

$R = 72 - 60 = 12$
by formula 2-4, $s^2 = \Sigma(x-\bar{x})^2/(n-1)$
$$= (204.50)/(9)$$
$$= 22.7$$
$$s = 4.8$$

Subtracting a constant from each score does not affect the spread of the scores or the measures of dispersion.

c.

x	x $-\bar{x}$	$(x-\bar{x})^2$
650	-65	4225
660	-55	3025
670	-45	2025
680	-35	1225
710	-5	025
730	15	225
740	25	625
770	55	3025
770	55	3025
770	55	3025
7150	0	20450

$$\bar{x} = (\Sigma x)/n$$
$$= 7150/10$$
$$= 715$$

$R = 770 - 650 = 120$
by formula 2-4, $s^2 = \Sigma(x-\bar{x})^2/(n-1)$
$$= (20450)/(9)$$
$$= 2272.2$$
$$s = 47.7$$

Multiplying each score by a constant also multiplies the distances between scores by that constant, and hence the measures of dispersion in the original units are multiplied by that constant (and the variance is multiplied by the square of that constant).

d.

x	x $-\bar{x}$	$(x-\bar{x})^2$
32.5	-3.25	10.5625
33.0	-2.75	7.5625
33.5	-2.25	5.0625
34.8	-1.75	3.0625
35.0	-0.25	0.0625
36.5	0.75	0.5625
37.0	1.25	1.5625
38.5	2.75	7.5625
38.5	2.75	7.5625
38.5	2.75	7.5625
357.5	0	51.1250

$$\bar{x} = (\Sigma x)/n$$
$$= 357.5/10$$
$$= 35.75$$

R = 38.5 - 32.5 = 6.0
by formula 2-4, $s^2 = \Sigma(x-\bar{x})^2/(n-1)$
$$= (51.1250)/(9)$$
$$= 5.68$$
$$s = 2.38$$
Dividing each score by a constant also divides the distances between scores by that constant, and hence the measures of dispersion in the original units are divided by that constant (and the variance is divided by the square of that constant).

23.

x	$x - \bar{x}$	$(x-\bar{x})^2$
65	-73.1	5343.61
66	-72.1	5198.41
67	-71.1	5055.21
68	-70.1	4914.01
71	-67.1	4502.41
73	-65.1	4238.01
77	-61.1	3733.21
77	-61.1	3733.21
77	-61.1	3733.21
740	601.9	362283.61
1381	0	402734.90

$\bar{x} = (\Sigma x)/n$
$= 1381/10$
$= 138.1$

R = 740 - 65 = 675
by formula 2-4, $s^2 = \Sigma(x-\bar{x})^2/(n-1)$
$$= (402734.90)/(9)$$
$$= 44748.3$$
$$s = 211.5$$
The inclusion of one outlier has changed the measures of dispersion considerably.

25. a. The limits 52.0 and 60.0 are 1 standard deviation from the mean. The empirical rule for bell-shaped data states that about 68% of the scores should fall within those limits.
   b. A distance of 8.0 is 2 standard deviations from the mean. The empirical rule for bell-shaped data states that about 95% of the scores should fall within those limits.
   c. The empirical rule for bell-shaped data states that about 99.7% of the scores should fall within 3 standard deviations of the mean. In this case, that would be within 3(4.0) = 12.0 of the mean of 56.0 -- i.e., from 44.0 to 68.0.

27. It is known from algebra that $1+2+3+\ldots+N = (N/2)(N+1)$. The mean of a population of those N values would then be $\mu = (\Sigma x)/N$
$$= [(N/2)(N+1)]/N$$
$$= (N+1)/2.$$
Exercise 26 gives The variance of the population 1,2,3,...,N as
$$\sigma^2 = (N^2 - 1)/12.$$
For N=165, the population of 1,2,3,...,164 has
$\mu = (N+1)/2 = (165+1)/2 = 83.0$
$\sigma^2 = (N^2 - 1)/12 = (165^2 - 1)/12 = 2268.7$
$\sigma = 47.6$
Exercise 21 of section 2-4 indicates that adding a constant to every score adds that constant to each measure of central tendency. Exercise 21 of section 2-5 indicates that adding a constant to every score does not affect the measures of dispersion. Since the population 18,19,20,...,182 is the population analyzed above but with 17 added to each score,
$\mu = 83.0 + 17 = 100.0$
$\sigma = 47.6$

29. Fahrenheit temperatures are converted to Celsius by $C = (5/9) \cdot F - 160/9$.
    Exercise 21 of section 2-4 indicates that multiplying each score by a constant multiplies the mean by that constant and that subtracting the same amount from each score subtracts that amount from the mean. The mean of the Celsius temperatures is therefore
$$\overline{x}_C = (5/9) \cdot \overline{x}_F - 160/9$$
$$= (5/9)(98.20) - 160/9$$
$$= 36.78$$
    Exercise 21 of section 2-5 indicates that multiplying each score by a constant multiplies the standard deviation by that constant and that subtracting the same amount from each score does not change the standard deviation of the scores. The standard deviation of the Celsius temperatures is therefore $s_C = (5/9) \cdot s_F$
$$= (5/9)(0.62)$$
$$= .34$$

31. $I = 3(\overline{x} - m)/s$
    $= 3(98.20 - 98.4)/.62$
    $= -.6/.62$
    $= -.968$
    The data are not *significantly* skewed.

2-6 Measures of Position

1. In general, $z = (x - \overline{x})/s$.
   a. $z_{98.6} = (98.6 - 98.20)/.62$
   $= .65$
   b. $z_{98.2} = (98.2 - 98.20)/.62$
   $= 0$
   c. $z_{100.0} = (100.0 - 98.20)/.62$
   $= 2.90$

3. $z = (x - \overline{x})/s$
   $z_{10.00} = (10.00 - 7.06)/5.32$
   $= .55$

5. $z = (x - \mu)/\sigma$
   a. $z_{70} = (70 - 63.6)/2.5$
   $= 2.56$
   b. Yes, the height is considered unusual since $2.56 > 2.00$.

7. $z = (x - \mu)/\sigma$
   a. $z_{10.00} = (10.00 - 22.83)/8.55$
   $= -1.50$
   b. No, the score is not considered unusual since $-2.00 < -1.50 < 2.00$.

9. In general $z = (x - \overline{x})/s$.
   a. $z_{53} = (53 - 50)/10$
   $= .30$
   b. $z_{53} = (53 - 50)/5$
   $= .60$
   The score in part b has the better relative position since $.60 > .30$.

11. In general, $z = (x - \bar{x})/s$.
    a. $z_{60} = (60 - 50)/5$
        $= 2.00$
    b. $z_{230} = (230 - 200)/10$
        $= 3.00$
    c. $z_{540} = (540 - 500)/15$
        $= 2.67$
    The score in part b has the highest relative position.

13. Let L = # of scores less than x
        n = total number of scores
    In general, the percentile of score x is $(L/n) \cdot 100$.
    The percentile of score 97.2 is $(8/106) \cdot 100 = 8$.

15. Let L = # of scores less than x
        n = total number of scores
    In general, the percentile of score x is $(L/n) \cdot 100$.
    The percentile of score 99.0 is $(98/106) \cdot 100 = 92$.

17. To find P_{80}, $L = (80/100) \cdot 106 = 84.8$ rounded up to 85.
    Since the 85th score is 98.7, $P_{80} = 98.7$.

19. To find $Q_3 = P_{75}$, $L = (75/100) \cdot 106 = 79.5$ rounded up to 80.
    Since the 80th score is 98.6, $Q_3 = 98.6$.

21. To find $D_3 = P_{30}$, $L = (30/100) \cdot 106 = 31.8$ rounded up to 32.
    Since the 32nd score is 97.9, $D_3 = 97.9$.

23. To find $D_7 = P_{70}$, $L = (70/100) \cdot 106 = 74.2$ rounded up to 75.
    Since the 75th score is 98.6, $D_7 = 98.6$.

25. Let L = # of scores less than x
        n = total number of scores
    In general, the percentile of score x is $(L/n) \cdot 100$.
    The percentile of score 120 is $(45/60) \cdot 100 = 75$.

NOTE: For exercises 25-36, refer to the following ordered list of the 60 film lengths.

	length (min.)		
1	72	99	111
2	82	100	114
3	88	100	115
4	90	100	117
5	90	101	119
6	91	102	120
7	92	103	121
8	92	104	123
9	93	104	123
0	93	104	125
1	94	104	129
2	94	105	133
3	94	105	134
4	95	106	139
5	96	106	144
6	96	107	155
7	96	108	159
8	97	108	160
9	98	110	168
0	98	111	193

27. Let L = # of scores less than x
        n = total number of scores
    In general, the percentile of score x is $(L/n) \cdot 100$.
    The percentile of score 100 is $(21/60) \cdot 100 = 35$.

29. To find P_{15}, $L = (15/100) \cdot 60 = 9$ -- a whole number.
    The mean of the 9th and 10th scores, $P_{15} = (93 + 93)/2 = 93$

31. To find P_{80}, $L = (80/100) \cdot 60 = 48$ -- a whole number.
    The mean of the 48th and 49th scores, $P_{80} = (123 + 123)2 = 123$.

33. To find $Q_1 = P_{25}$, $L = (25/100) \cdot 60 = 15$ -- a whole number.
    The mean of the 15th and 16th scores, $Q_1 = (96 + 96)/2 = 96$.

35. To find $D_9 = P_{90}$, $L = (90/100) \cdot 60 = 54$ -- a whole number.
    The mean of the 54th and 55th scores, $D_9 = (139 + 144)/2 = 141.5$.

37. a. The interquartile range is $Q_3 - Q_1$.
    For $Q_3 = P_{75}$, $L = (75/100) \cdot 106 = 79.5$ rounded up to 80.
      Since the 80th score is 98.6, $Q_3 = 98.6$.
    For $Q_1 = P_{25}$, $L = (25/100) \cdot 106 = 26.5$ rounded up to 27.
      Since the 27th score is 97.8, $Q_1 = 97.8$.
    The interquartile range is $98.6 - 97.8 = 0.8$

  b. The midquartile is $(Q_1 + Q_3)/2 = (98.6 + 97.8)/2$
    $$= 98.2$$

  c. The 10-90 percentile range is $P_{90} - P_{10}$.
    For P_{90}, $L = (90/100) \cdot 106 = 95.4$ rounded up to 96.
      Since the 96th score is 98.8, $P_{90} = 98.8$.
    For P_{10}, $L = (10/100) \cdot 106 = 10.6$ rounded up to 11.
      Since the 11th score is 97.3, $P_{10} = 97.3$.
    The 10-90 percentile range is $98.8 - 97.3 = 1.5$

  d. Yes, $Q_2 = P_{50}$ by definition. They are always equal.

  e. For $Q_2 = P_{50}$, $L = (50/100) \cdot 106 = 53$ exactly;
    we use the mean of the 53rd and 54th score.
    $Q_2 = (98.4 + 98.4)/2 = 98.4$.
    In this case $Q_2 \neq (Q_1 + Q_3)/2$, demonstrating that the median does not necessarily equal the midquartile.

39. Unusual values are those more than two standard deviations from the mean. In this case, that would be any values such that $x < \bar{x} - 2 \cdot s$ or $x > \bar{x} + 2 \cdot s$

$$x < 98.20 - 2(.62) \quad \text{or} \quad x > 98.20 + 2(.62)$$
$$x < 96.96 \quad \text{or} \quad x > 99.44$$

NOTE: The practical interpretation of this is that a temperature below 96.96 or above 99.44 is unusual for a well person. The logical conclusion is that such a person either is not well (typically arrived at in the presence of other symptoms) or has an unusual "normal" temperature.

2-7   Exploratory Data Analysis

1. 406 406 407 408
   410 419 419 419 419
   421 423 424 426 426
   430 438 438

3. First go through the numbers in the order they appear and place the one's digit behind the appropriate ten's digit. Going through the numbers by rows produces the following.

```
Temperature
   6 | 84
   7 | 7659
   8 | 046519233441400
   9 | 3473024021
```

Now order the one's digits within each row to produce the following final form.

```
Temperature
   6 | 48
   7 | 5679
   8 | 000112334444569
   9 | 0012233447
```

5. Make the stem accurate to tenths, and make two rows for each tenth -- one for low hundredth digits (0-4) and one for high hundredth digits (5-9). The final form, with the hundredth digits in order, is given below.

```
Weight (grams)
   5.4 | 9
   5.5 | 2333
   5.5 | 677777888889999
   5.6 | 00000001222333
   5.6 | 5666778
   5.7 | 01123334
   5.7 |
   5.8 | 4
```

7. The 10 scores are given in order in the text.
   The minimum score is .02.
   The maximum score is .19.
   The median score is .095.
   The left hinge is .08, median of .02 .04 .08 .08 .09 .095.
   The right hinge is .11, median of .095 .10 .10 .12 .13 .19.

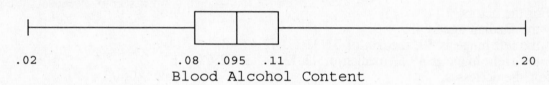

```
    .02                 .08 .095 .11                              .20
                   Blood Alcohol Content
```

9. Arranged in order, the 20 scores are
   ```
    70  130  138  142  157  157  159  162  164  170
   173  173  175  180  181  183  190  193  195  198
   ```
   The minimum score is 70.
   The maximum score is 198.
   The median score is 171.5.
   The left hinge is 157, median of 70 130 ... 164 170 171.5.
   The right hinge is 181, median of 171.5 173 173 ... 195 198.

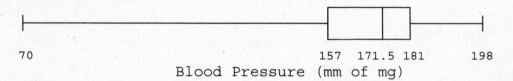

```
    70                      157  171.5 181      198
              Blood Pressure (mm of mg)
```

11. Arranged in order, the 30 scores are
   ```
     2   4   7   7   9   9  10  12  14  15  18  21  22  25  26
    27  29  30  31  31  32  34  35  36  36  38  40  41  43  45
   ```
   The minimum score is 2.
   The maximum score is 45.
   The median score is 26.5.
   The left hinge is 13, median of 2 4 ... 25 26 26.5.
   The right hinge is 34.5, median of 26.5 27 29 ... 43 45.

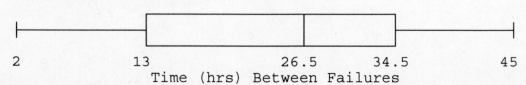

```
    2          13          26.5  34.5         45
           Time (hrs) Between Failures
```

13. Lottery numbers should be uniformly distributed, in which case the hinges and the center line in the boxplot would divide the figure into four equal lengths. The given boxplot indicates a concentration of numbers near the middle, with very low and very high numbers occurring less frequently.

15. a. The final form of the back-to-back stem-and-leaf plot is given below. NOTE: This is another example of adapting a standard visual form in order to better communicate the data (see exercise 2.3 #21). While such decisions are arbitrary, we choose to display "outward" from the central stem but to keep the actor's ages in increasing order from left to right.

```
    actor's age    actress' age
                 2│146667
    122357899    3│00113344455778
 0012233456788   4│11129
         1566    5│
          012    6│011
            6    7│4
                 8│0
```

b. For the actors,
   the median age is 43.
   the left hinge is 39, median of 31 31 ... 42 43 43.
   the right hinge is 49.5, median of 43 43 44 ... 62 67.
For the actresses,
   the median age is 34.5.
   the left hinge is 30.5, median of 21 24 ... 34 34 34.5.
   the right hinge is 41, median of 34.5 35 35 ... 74 80.
The boxplots, using the same scale, are shown below.

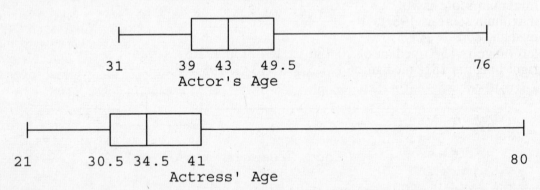

c. Female Oscar winners tend to be younger than male Oscar winners. Assuming that acting ability doesn't peak differently for females and males, the data may reveal a difference in the standards by which way females and males are judged.

Review Exercises

1. Arranged in order, the scores are:
   4.10 4.42 4.95 5.01 7.24 7.51 7.98 8.60 11.53 12.30 14.59 14.67
   preliminary values: $n = 12$, $\Sigma x = 102.90$, $\Sigma x^2 = 1044.7150$
   a. $\bar{x} = (\Sigma x)/n = (102.90)/12 = 8.575$
   b. $m = (7.51 + 7.98)/2 = 7.745$
   c. $M = $ [none]
   d. $m.r. = (4.10 + 14.67)/2 = 9.385$
   e. $R = 14.67 - 4.10 = 10.57$
   f. $s^2 = [n(\Sigma x^2) - (\Sigma x)^2]/[n(n-1)]$
   $\quad = [12(1044.7150) - (102.90)^2]/[12(11)]$
   $\quad = (1948.1700)/132$
   $\quad = 14.759$
   g. $s = 3.842$

3. The relative frequency for each class is obtained by dividing that class' frequency by
   $n = \Sigma f = 300$.

class boundaries	relative frequency
3.5 - 4.5	.49
4.5 - 5.5	.27
5.5 - 6.5	.09
6.5 - 7.5	.05
7.5 - 11.5	.10
	1.00

   The relative frequency histogram below is constructed from the above relative frequency
   distribution. NOTE: The purpose of a histogram is to show the relative amounts in each class
   using areas of bars -- i.e., the proportion of the total area in each bar should represent the
   proportion of the sample in that class. Since the last class is 4 times as wide as the other
   classes, its bar must be made 1/4 of its normal height in order to preserve proper proportions.
   The <u>area</u> of the last bar should be about equal to the <u>area</u> of the 5.5-6.5 bar and twice the <u>area</u>
   of the 6.5-7.5 bar; this conveys the information that 10% of the data were spread over the
   years 8,9,10,11 -- assumed to be about 2.5% of the data for each of those years.

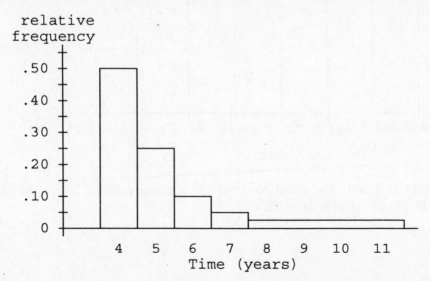

5. Subtracting the smallest value from the largest value and dividing by the desired number of classes determines that the class width must be at least $(1128 - 235)/9 = 99.22$. Rounding up to a convenient number, we choose 100. The exercise specifies that the first lower class limit is to be 235, the smallest data value. NOTE: In practice, the first lower class limit and the class width many be any convenient values. Since there are 60 seconds in a minute, for example, it might be reasonable to start at 180 or 210 and use a class width of 90 or 120.

time (seconds)	frequency
235 - 334	4
335 - 434	9
435 - 534	11
535 - 634	9
635 - 734	9
735 - 834	6
835 - 934	8
935 - 1034	2
1035 - 1134	2
	60

7. frequency

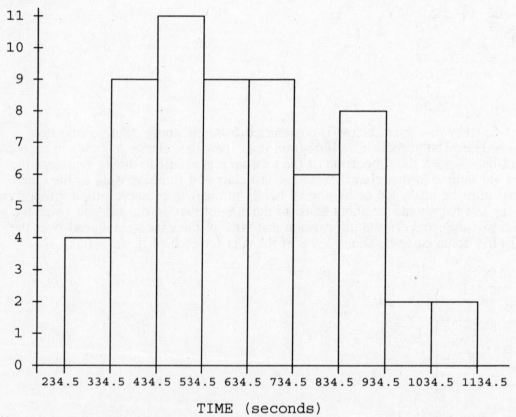

TIME (seconds)

9. According to the range rule of thumb, the standard deviation is usually about 1/4 of the range -- i.e., we estimate s to be about $(1/4)(893) = 223.25$.

11. Use the hundred's digits for the stem and the last two digits -- with spaces between values -- for the leaves.

first pass, by columns

```
 2 | 92 40 35
 3 | 25 37 35 78 63 96 96 45
 4 | 48 43 04 57 47 95 94 20 74 83
 5 | 87 14 06 40 03 64 52
 6 | 26 70 15 88 09 25 76 70 93 66 27
 7 | 56 23 00 94 93 04 78 48
 8 | 61 71 20 53 52 60 62
 9 | 91 29 15
10 | 23 70
11 | 28
```

final form

```
 2 | 35 40 92
 3 | 25 35 37 45 63 78 96 96
 4 | 04 20 43 47 48 57 74 83 94 95
 5 | 03 06 14 40 52 64 87
 6 | 09 15 25 26 27 66 70 70 76 88 93
 7 | 00 04 23 48 56 78 93 94
 8 | 20 52 53 60 61 62 71
 9 | 15 29 91
10 | 23 70
11 | 28
```

13. Arranged in order, the scores are: 8 8 12 16 18 19 22 25
   preliminary values: $n = 8$, $\Sigma x = 128$, $\Sigma x^2 = 2322$
   a. $\bar{x} = (\Sigma x)/n = (128)/8 = 16.0$
   b. $m = (16 + 18)/2 = 17.0$
   c. $M = 8$
   d. m.r. $= (8 + 25)/2 = 16.5$
   e. $R = 25 - 8 = 17$
   f. $s^2 = [n(\Sigma x^2) - (\Sigma x)^2]/[n(n-1)]$
   $\quad = [8(2322) - (128)^2]/[8(7)]$
   $\quad = (2192)/56$
   $\quad = 39.1$
   g. $s = 6.3$

15. Arranged in order, the scores are: 2 4 4 10 12 15 17 17 22 25 27 28 29 29 29 30 31 32 32
   preliminary values: $n = 19$, $\Sigma x = 395$, $\Sigma x^2 = 10137$
   a. $\bar{x} = (\Sigma x)/n = (395)/19 = 20.8$
   b. $m = 25.0$
   c. $M = 29$
   d. m.r. $= (2 + 32)/2 = 17.0$
   e. $R = 32 - 2 = 30$
   f. $s^2 = [n(\Sigma x^2) - (\Sigma x)^2]/[n(n-1)]$
   $\quad = [19(10137) - (395)^2]/[19(18)]$
   $\quad = (36578)/342$
   $\quad = 107.0$
   g. $s = 10.3$

Chapter 3

Probability

3-2 Fundamentals

1. Since $0 \leq P(A) \leq 1$ is always true, the following values less than 0 or greater than 1 cannot be probabilities.
   values less than 0: -0.2
   values greater than 1: 4/3   1.001   2 $\sqrt{2}$

3. let C = the selected student cheated
   P(C) = 1162/3630 = .320

5. let O = a selected person has group O blood
   P(O) = 36/80 = .45

7. let V = a selected 12-24 year old voted
   P(V) = (9,230,000)/(25,569,000) = .361

9. There were 288 + 962 = 1250 households surveyed.
   let C = a selected household has a computer
   P(C) = 288/1250 = .230

11. let S = a selected 18-25 year old smokes
    P(S) = 237/600 = .395

13. let A = a selected 20-24 year old has an accident
    P(A) = 136/400 = .34
    NOTE: This is properly an estimate of the probability that such a person <u>had</u> an accident last year. Using this to estimate the probability that such a person <u>will have</u> an accident next year is frequently done but requires the additional assumption that relevant conditions (speed limits, weather, alcohol laws, the economy, etc., etc.) remain relatively unchanged.

15. There were 70 + 711 persons tested.
    let D = a selected user becomes drowsy
    P(D) = 70/781 = .0896

17. There were 90 + 80 + 20 + 10 = 200 people surveyed.
    let E = a selected person has group AB blood
    P(E) = 10/200 = .05

19. There were 10,589 + 636 + 22,551 + 963 = 34,739 convictions.
    There were 636 + 963 = 1599 DWI convictions.
    let D = a selected conviction is for DWI
    P(D) = 1599/34739 = .0460

21. a. The 4 equally likely outcomes are: BB BG GB GG.
    b. P(2 girls) = 1/4 = .25
    c. P(1 of each sex) = 2/4 = .5

23. a. The 8 equally likely outcomes are: CCC   CNN
      let C = correct   CCN   NCN
      let N = not correct   CNC   NNC
                                        NCC   NNN
    b. P(3 correct) = 1/8 = .125   [CCC above]
    c. P(0 correct) = 1/8 = .125   [NNN above]
    d. P(at least 2 correct) = 4/8 = .5 [column 1 above]

25. There were 20 managers surveyed.
    There were 12 managers with values of 2.1 or higher.
    let M = a selected manager so spends more than 2 hours
    P(M) = 12/20 = .6

27. No matter where the two flies land it is possible to cut the orange in half to include both flies on the same half. Since this is a certainty, the probability is 1. [Compare the orange to a globe. Turn the orange so the spot where fly A lands corresponds to, say, New York City. Consider all the circles of longitude. Wherever fly B lands it is possible to slice the globe in half along some circle of longitude that places fly A and fly B on the same half.] NOTE: If the orange is marked into two predesignated halves before the flies land, the problem is different; once fly A lands, fly B has 1/2 a chance of landing on the same half. If both flies are to land on a specified one of the two predesignated halves, the problem is different still; fly A has 1/2 a chance of landing on the specified half, and only 1/2 the time will fly B pick the same half -- the final answer would then be 1/2 of 1/2, or 1/4.

29. let W = winning by selecting the correct slot on the wheel
    P(W) = 1/38
    P($\overline{W}$) = 37/38
    a. odds against = P($\overline{W}$)/P(W) = (37/38)(1/38) = 37/1 or 37:1
    b. odds in favor = 1:37
    NOTE: If the odds against are 37:1, then a win gives $37 in winnings to a person putting up $1. Since P(W) = 1/38, someone who plays 38 times expects to win one time in 38. That one win will provide the $37 to cover the 37 losses, and the game is fair. In practice, a casino will offer odds of, say, 35:1 so that it will be likely to earn a profit.

3-3 Addition Rule

1. a. No, a student may regularly attend statistics class and own a computer.
    b. No, a student may have blond hair and brown eyes.
    c. Yes, a course cannot be required and be an elective at the same time (i.e., in the same context).

3. a. P($\overline{A}$) = 1 - P(A)
             = 1 - .45
             = .55
    b. P(girl) = P($\overline{boy}$)
               = 1 - P(boy)
               = 1 - .513
               = .487

5. It may be helpful to make a chart like the one given at the right.

AGE	RESPOND?		
	Yes	No	
18-21	73	11	84
22-29	255	20	275
	328	31	359

let A = a person is 18-21
  N = a person does not respond

There are two correct approaches.

* Use broad categories and allow for double-counting.

$$P(A \text{ or } N) = P(A) + P(N) - P(A \text{ and } N)$$
$$= 84/359 + 31/359 - 11/359$$
$$= 104/359$$
$$= .290$$

* Use individual mutually exclusive categories that involve no double-counting. NOTE: For simplicity in this problem, we use B for "a person is 22-29" and AY for "A and Y."

$$P(A \text{ or } N) = P(AY \text{ or } AN \text{ or } BN)$$
$$= P(AY) + P(AN) + P(BN)$$
$$= 73/359 + 11/359 + 20/359$$
$$= 104/359$$
$$= .290$$

NOTE: In general, using broad categories and allowing for double-counting is a "more powerful" technique that "lets the formula do the work" and requires less detailed analysis by the solver. Except when such detailed analysis is instructive, the solutions in this manual utilize the first approach.

7. It may be helpful to make a chart like the one given at the right.

SEX	HARASSMENT?		
	Yes	No	
Male	240	380	620
Female	180	200	380
	420	580	1000

let N = a person says no harassment
$$P(N) = 580/1000$$
$$= .58$$

9. It may be helpful to make a chart like the one given at the right.

LOCATION	SMOKING?		
	Yes	No	
Aisle	16	64	80
Off-A	24	96	120
	40	160	200

let A = an aisle seat
  Y = a seat with smoking permitted
$$P(A \text{ or } Y) = P(A) + P(Y) - P(A \text{ and } Y)$$
$$= 80/200 + 40/200 - 16/200$$
$$= 104/200$$
$$= .52$$

11. It may be helpful to make a chart like the one given at the right.

STATUS	SMOKE?		
	Yes	No	
Married	54	146	200
Divorced	38	62	100
Alw Single	11	39	50
	103	247	350

let D = a person is divorced
  Y = a person smokes
$$P(D \text{ or } Y) = P(D) + P(Y) - P(D \text{ and } Y)$$
$$= 100/350 + 103/350 - 38/350$$
$$= 165/350$$
$$= .471$$

13. The total number in the survey is 95,277.
    let x = the age of a person in the survey
    $P(x<5 \text{ or } x>74) = P(x<5) + P(x>74)$ [mutually exclusive]
                   $= 3843/95277 + 16800/95277$
                   $= 20643/95277$
                   $= .217$

15. The total number in the survey is 95,277.
    There are 55,245 younger than 45.
    There are 50,433 between 25 and 74 inclusive.
    let x = the age of a person in the survey
    $P(x<45 \text{ or } 25\leq x\leq74) = P(x<45) + P(25\leq x\leq74) - P(25\leq x<45)$
                   $= 55245/95277 + 50433/95277 - 27201/95277$
                   $= 78477/95277$
                   $= .824$
    NOTE: This is of the form $P(A \text{ or } B) = P(A) + P(B) - P(A \text{ and } B)$
                   where, $A = x<45$
                           $B = 25\leq x\leq74$
                   $A \text{ and } B = x<45 \text{ and } 25\leq x\leq74$
                           $= 25\leq x<45$

17. It may be helpful to make a chart like
    the one given at the right.

    $P(A \text{ or } B) = P(A) + P(B)$
              $= 40/100 + 10/100$
              $= 50/100$
              $= .5$

		RH FACTOR		
		+	-	
	A	35	5	40
	B	8	2	10
GROUP	AB	4	1	5
	O	39	6	45
		86	14	100

19. Refer to exercise 17.
    $P(A \text{ or } -) = P(A) + P(-) - P(A \text{ and } -)$
              $= 40/100 + 14/100 - 5/100$
              $= 49/100$
              $= .49$

21. Refer to exercise 17.
    $P(\text{not } AB) = 1 - P(AB)$
              $= 1 - 5/100$
              $= 1 - .05$
              $= .95$

23. Refer to exercise 17.
    $P(O \text{ or } -) = P(O) + P(-) - P(O \text{ and } -)$
              $= 45/100 + 14/100 - 6/100$
              $= 53/100$
              $= .53$

25. a. $P(A \text{ or } B) = P(A) + P(B) - P(A \text{ and } B)$
       $1/3 = P(A) + 1/4 - 1/5$
       $.333 = P(A) + .250 - .200$
       $.333 = P(A) + .050$
       $.283 = P(A)$

   b. for mutually exclusive events, $P(A \text{ or } B) = P(A) + P(B)$
$$= .4 + .5$$
$$= .9$$
   c. if A and B are not mutually exclusive, $P(A \text{ and } B) > 0$
   and so, $P(A \text{ or } B) = P(A) + P(B) - P(A \text{ and } B)$
$$= .4 + .5 - P(A \text{ and } B)$$
$$= .9 - P(A \text{ and } B)$$
$$= \text{something less than } .9$$
   NOTE: More precisely, whenever A and B are not mutually exclusive it must be true that
$$0 < P(A \text{ and } B) \leq \min[P(A),P(B)]$$
   here, $0 < P(A \text{ and } B) \leq .4$
   and so, $.5 \leq P(A \text{ or } B) < .9$

27. If the exclusive *or* is used instead of the inclusive *or*, then the double-counted probability must be completely removed (i.e., must be subtracted twice) and the formula becomes
$$P(A \text{ or } B) = P(A) + P(B) - 2 \cdot P(A \text{ and } B)$$

3-4 Multiplication Rule

1. a. Dependent, since removing the first watch changes the probabilities of getting a defective watch on the second selection.
   b. Independent, since getting the first question right or wrong does not affect the chance of guessing correctly on the second question.

3. let N = the reservation is for a non-smoking table
   $P(N) = .65$, for all selections
   $P(N_1 \text{ and } N_2 \text{ and } N_3 \text{ and } N_4) = P(N_1) \cdot P(N_2) \cdot P(N_3) \cdot P(N_4)$
$$= (.65)(.65)(.65)(.65)$$
$$= .179$$

5. let H = the derailment was caused by human error
   $P(H) = .24$, for all selections
   $P(D_1 \text{ and } D_2 \text{ and } D_3 \text{ and } D_4 \text{ and } D_5) = P(D_1) \cdot P(D_2) \cdot P(D_3) \cdot P(D_4) \cdot P(D_5)$
$$= (.24)(.24)(.24)(.24)(.24)$$
$$= .000796$$

7. let D = the disk selected is defective
   $P(D) = 10/30$ for the first selection only
   $P(D_1 \text{ and } D_2 \text{ and } D_3 \text{ and } D_4) = P(D_1) \cdot P(D_2 \mid D_1)...$
$$= (10/30)(9/29)(8/28)(7/27)$$
$$= .00766$$

9. let W = a new car buyer is a woman
   a. $P(W) = .5$, for each sale
   $P(W_1 \text{ and } W_2 \text{ and } W_3 \text{ and } W_4) = P(W_1) \cdot P(W_2) \cdot P(W_3) \cdot P(W_4)$
$$= (.5)(.5)(.5)(.5)$$
$$= .0625$$
   b. $P(W) = .6$, for each sale
   $P(W_1 \text{ and } W_2 \text{ and } W_3 \text{ and } W_4) = P(W_1) \cdot P(W_2) \cdot P(W_3) \cdot P(W_4)$
$$= (.6)(.6)(.6)(.6)$$
$$= .130$$

11. let J = a person is born on July 4
 P(J) = 1/365, for all random selections
 P(J$_1$ and J$_2$ and J$_3$) = P(J$_1$)·P(J$_2$)·P(J$_3$)
 = (1/365)(1/365)(1/365)
 = .0000000206
NOTE: This is, as the problem requests, the probability of 3 randomly chosen people being born on July 4. There might be special circumstances that change that probability for the family in question -- suppose, for example, the husband is a migrant worker who is away all summer and returns to his wife each year around October 4 (i.e., 9 months before July 4).

13. let A = a death is accidental
 P(A) = .0478, for all random selections
 P(A$_1$ and A$_2$ and A$_3$ and A$_4$ and A$_5$) = P(A$_1$)·P(A$_2$)·P(A$_3$)·P(A$_4$)·P(A$_5$)
 = (.0478)^5
 = .000000250
NOTE: This is, as the problem requests, the probability of 4 randomly chosen deaths being accidental. The deaths investigated by the Baltimore detective, however, involved special circumstances and were not chosen at random.

15. let V = an eligible voter actually votes
 P(V) = .6, for each selection
 P(V$_1$ and V$_2$ and...and V$_9$ and V$_{10}$) = (.6)^{10}
 = .00605

17. let N = the flight from New York is on time
 S = the flight from St. Louis is on time
 P(N) = .80
 P(S) = .60
 P(N and S) = P(N)·P(S)   [assuming independence]
 = (.8)(.6)
 = .48

19. let F = a selection has faulty brakes
 P(F) = 3/12 for the first selection (if the owner is correct)
 P(F$_1$ and F$_2$ and F$_3$) = P(F$_1$)·P(F$_2$ | F$_1$)·P(F$_3$ | F$_1$ and F$_2$)
 = (3/12)·(2/11)·(1/10)
 = .00455

21. let L = losing when betting on #7
 P(L) = 37/38, each time
 P(L$_1$ and L$_2$ and...and L$_{37}$ and L$_{38}$) = [P(L)]^{38}
 = [37/38]^{38}
 = .363

23. let N = no reply is received from the coupon
 P(N) = .8, for each coupon
 P(N$_1$ and N$_2$ and N$_3$ and N$_4$ and N$_5$ and N$_6$) = [P(N)]^6
 = [.8]^6
 = .262

25. let N = an individual tests negative
$P(N)$ = .985, for each person
P(group is positive) = 1 - P(the group is negative)
$$= 1 - P(N_1 \text{ and } N_2 \text{ and } N_3 \text{ and } N_4 \text{ and } N_5)$$
$$= 1 - (.985)(.985)(.985)(.985)(.985)$$
$$= 1 - .9272$$
$$= .0728$$

27. let F = a component fails
$P(F)$ = .181, for each component
P(circuit works) = 1 - P(circuit fails)
$$= 1 - P(F_1 \text{ and } F_2 \text{ and } F_3 \text{ and } F_4)$$
$$= 1 - (.181)(.181)(.181)(.181)$$
$$= 1 - .001$$
$$= .999$$

29. Refer to the table reproduced from the text and given at the right.

	CRIME			
	Hom	Rob	Asl	
stranger	12	379	727	1118
CRIMINAL frnd/rel	39	106	642	787
unknown	18	20	57	95
	69	505	1426	2000

let H,R A stand for homicide robbery, assault
let S,F,U stand for stranger, friend/relative, unknown
a. This problem may be done either of two ways. NOTE: This is a conditional probability, given that the crime was a robbery. When using the first method below, the robbery column becomes the relevant sample space and is the only column used to answer the question.
  * reading directly from the table
    $P(S \mid R)$ = 379/505
    $$= .750$$
  * using the formula
    $P(S \mid R)$ = P(S and R)/P(R)
    $$= (379/2000)/(505/2000)$$
    $$= 379/505$$
    $$= .750$$
b. P(R and S) = 379/2000
    $$= .1895$$
c. P(R or U) = P(R) + P(U) - P(R and U)
    $$= (505/2000) + (95/2000) - (20/2000)$$
    $$= 580/2000$$
    $$= .290$$

31. Refer to the table and notation for exercise 29. NOTE: This is a good exercise to verify that one can distinguish whether a word problem is asking an "or" or "and" or "given" question.
a. P(A or S) = P(A) + P(S) - P(A and S)
    $$= (1426/2000) + (1118/2000) - (727/2000)$$
    $$= 1817/2000$$
    $$= .9085$$
b. P(A and S) = 727/2000
    $$= .3635$$

c. * reading directly from the table
$$P(A \mid S) = 727/1118$$
$$= .650$$
* using the formula
$$P(A \mid S) = P(A \text{ and } S)/P(S)$$
$$= (727/2000)/(1118/2000)$$
$$= 727/1118$$
$$= .650$$

33. a. let D = a birthday is different from any yet selected
$P(D_1) = 366/366$ NOTE: With nothing to match, it <u>must</u> be different.
$P(D_2 \mid D_1) = 365/366$
$P(D_3 \mid D_1 \text{ and } D_2) = 364/366$
...
$P(D_{25} \mid D_1 \text{ and } D_2 \text{ and}...\text{and } D_{24}) = 342/366$
$$P(\text{no match}) = P(D_1 \text{ and } D_2...D_{25})$$
$$= (366/366) \cdot (365/366) \cdot (364/366) \cdots (342/366)$$
$$= .432$$
NOTE: Programs to perform this calculation can be constructed as follows.

*** in BASIC	*** in Minitab
10 LET P=1	*MTB > LET K1=1*
20 FOR K=1 TO 24	*MTB > LET K2=1*
30 LET P=P(366-K)/366*	*MTB > STORE*
40 NEXT K	*STOR> LET K1=K1*(366-K2)/366*
50 PRINT P	*STOR> LET K2=K2+1*
60 END	*STOR> END*
	MTB > NOECHO
	MTB > EXECUTE 24
	MTB > PRINT K1

The STORE command in Minitab allows the creation of a file of commands. The EXECUTE N command instructs Minitab to execute the stored file N times. The NOECHO command suppresses intermediate printout. When February 29 is ignored and a 365-day year is used, the answer is .431.

b. P(at least one match) = 1 - P(no match)
$$= 1 - .432$$
$$= .568$$

35. The following English/logic facts are used in this exercise.
* not (A or B) = not A and not B
Is your sister either artistic or bright?
No? Then she is not artistic and she is not bright.
* not (A and B) = not A or not B
Is your brother artistic and bright?
No? Then either he is not artistic or he is not bright.
a. $P(\overline{A \text{ or } B}) = P(\overline{A} \text{ and } \overline{B})$ [from the first fact above]
    or
$P(\overline{A \text{ or } B}) = 1 - P(A \text{ or } B)$ [rule of complementary events]
b. $P(\overline{A} \text{ or } \overline{B}) = 1 - P(A \text{ and } B)$ [from the second fact above]
c. They are different: (a) gives the complement of "A or B"
                    while (b) gives the complement of "A and B."

37. This is problem can be done by two different methods. In either case,
    let T = getting a 10
       C = getting a club
    * consider the sample space
    The first card could be any of 52 cards; for each first card, there are 51 possible second cards. This makes a total of $52 \cdot 51 = 2652$ equally likely outcomes in the sample space. How many of them are T_1C_2?
    The 10's of hearts, diamonds and spades can be paired with any of the 13 clubs for a total of $3 \cdot 13 = 39$ favorable possibilities. The 10 of clubs can only be paired with any of the remaining 12 members of that suit for a total of 12 favorable possibilities. Since there are $39 + 12 = 51$ favorable possibilities among the equally likely outcomes,
$$P(T_1C_2) = 51/2652$$
$$= .0192$$

    * use the formulas
    Let Tc and To represent the 10 of clubs and the 10 of any other suit respectively. Break T_1C_2 into mutually exclusive parts so the probability can be found by adding and without having to consider double-counting.
$$
\begin{aligned}
P(T_1C_2) &= P(Tc_1C_2 \text{ or } To_1C_2) \\
&= P(Tc_1C_2) + P(To_1C_2) \\
&= (1/52)(12/51) + (3/52)(13/51) \\
&= 12/2652 + 39/2652 \\
&= 51/2652 \\
&= .0192
\end{aligned}
$$

3-5 Probabilities Through Simulations

1. Following the example in the text and using even digits to represent boys and odd digits to represent girls produces the following list of $50 \cdot 3 = 150$ simulated births. A slash follows the birth of each girl, and the family size necessary to produce a girl is the number of children since the last the slash. NOTE: The preceding system is only one possible method for converting the random numbers to sex of children. One could have, for example, used the digits 0-4 to represent boys and 5-9 to represent girls to produce a different list. The end result, however, should be similar for any valid simulation.

```
BBBG/G/G/G/BBBBBBBG/G/BG/BBG/BG/G/BG/G/G/G/G/BBBG/G/G/
G/BG/BG/BG/G/G/G/BG/BBG/BBBBG/BG/G/G/G/BG/BBBG/BG/G/BB
BG/BG/BBG/BBG/G/G/BG/G/G/BBG/BBG/BBBG/G/G/G/G/BG/BG/BB
G/G/BG/G/G/BBBG/G/BG/BG/G/G/G/BBG/BBG/G/G/BG/G/G/G/G/
G/G/G/BG/G/BG/B (the final B is not used)
```

The 82 x values (necessary family sizes) generated are:

```
4 1 1 1 8 1 2 3 2 1 2 1 1 1 1 4 1 1 1 2 2 2 1 1 1 2 3 5
2 1 1 1 2 4 2 1 4 2 3 3 1 1 2 1 1 3 3 4 1 1 1 1 2 2 3 1
2 1 1 4 1 2 2 1 1 1 3 3 1 1 2 1 1 1 1 1 1 1 2 1 2
```

Using the sample mean from the simulation to estimate the true average number of births necessary to get a girl, $\bar{x} = \Sigma x/n = 149/82 = 1.8$.

NOTE: Some problems are so complicated that the true answer will probably never be known and the simulated answer (or the average of several simulated answers) must be assumed to be close enough to allow the research to proceed. In this instance, the true average number may be found from probability theory to be 2.0. It is a weighted average of the possible necessary family sizes 1,2,3... where the probabilities are the weights and

$P(x=1) = P(G) = .5$

$P(x=2) = P(BG) = P(B_1) \cdot P(G_2 \mid B_1) = (.5)(.5) = (.5)^2$

$P(x=3) = P(BBG) = (.5)(.5)(.5) = (.5)^3$

...

in general, $P(x) = (.5)^x$ and $\Sigma P(x) = 1$

The true average is the weighted mean $\mu = (\Sigma x \cdot w)/(\Sigma w)$
$$= (\Sigma x \cdot P(x))/(\Sigma P(x))$$
$$= \Sigma x \cdot P(x)$$
$$= \Sigma x \cdot (.5)^x \quad \text{for the summation } x=1,2,3...$$

The more mathematically inclined may wish to follow the proof given below that $\mu = 2$.

Replacing x with y+1,

$\mu = \Sigma(y+1) \cdot (.5)^{y+1}$ for the summation y=0,1,2,3...

$\quad = 1 \cdot (.5)^1 + \Sigma(y+1) \cdot (.5)^{y+1}$ for the summation y=1,1,2,3...

$\quad = .5 \quad\quad + \Sigma y \cdot (.5)^{y+1} + \Sigma 1 \cdot (.5)^{y+1}$ for the summation y=.2,3...

$\quad = .5 \quad\quad + (.5) \cdot \Sigma y \cdot (.5)^y + (.5) \cdot \Sigma(.5)^y$ for the summation y=1,2,3...

$\quad = .5 \quad\quad + (.5) \cdot \mu \quad\quad + (.5) \cdot 1$

Which produces the result

$\mu = .5 + .5\mu + .5$

$.5\mu = 1$

$\mu = 2$

3. Using 0 to represent a good chip and 1,2,...,9 to represent a defective one produces the following list of 13 simulated numbers of chips necessary to get a good one:

1 1 1 44 6 1 3 15 1 12 10 20 20

NOTE: The last 15 digits could not be used.

Using the sample mean from the simulation to estimate the true average number of chips necessary to get a good one, $\bar{x} = \Sigma x/n = 135/13 = 10.4$. NOTE: The true average number may be found from probability theory to be 10.0.

5. Using 0 to represent a good chip and 1,2,...,9 to represent a defective one produces the following list of 6 simulated numbers of chips necessary to get 2 good ones:

2 45 7 18 13 30

NOTE: The last 35 digits could not be used.

Using the sample mean from the simulation to estimate the true average number of chips necessary to get a good one, $\bar{x} = \Sigma x/n = 115/6 = 19.2$. NOTE: The true average number may be found from probability theory to be 20.0

7. Using 1,2,3,4,5,6 to represent those die results and ignoring 7,8,9,0 produces the following list of 18 simulated numbers of tosses necessary to get a 6:

4 2 1 3 1 3 1 9 6 3 7 6 2 4 22 1 3 16

Using the sample mean from the simulation to estimate the true average number of tosses necessary to get a 6, $\bar{x} = \Sigma x/n = 94/18 = 5.2$. NOTE: The true average number may be found from probability theory to be 6.0.

9. Using even digits to represent boys and odd digits to represent girls produces the following list of 50 simulated numbers of girls per 3-child family:

```
0 3 1 0 1 2 1 2 2 3 0 3 2 1 3 2 1 0 1 3 2 0 2 1 1
1 1 3 2 1 1 1 2 3 1 1 2 3 0 2 2 3 1 1 2 3 3 3 2 1
```

Using the sample mean from the simulation to estimate the true average number of girls in a 3-child family, $\bar{x} = \Sigma x/n = 82/50 = 1.64$. NOTE: The true average number may be found from probability theory to be 1.5.

Using the sample standard deviation mean from the simulation to estimate the true standard deviation among the numbers of girls in a 3-child family,

$$s^2 = [n(\Sigma x^2) - (\Sigma x)^2]/[n(n-1)]$$
$$= [50(182) - (82)^2]/[50(49)]$$
$$= 2376/2450$$
$$= .970$$
$$s = .98$$

NOTE: The true standard deviation among the numbers may be found from probability theory to be .866

3-6 Counting

1. $7! = 7 \cdot 6 \cdot 5 \cdot 4 \cdot 3 \cdot 2 \cdot 1$
$$= 5040$$

3. $(70!)/(68!) = (70 \cdot 69 \cdot 68 \cdot 67 \cdot 66 \cdots)/(68 \cdot 67 \cdot 66 \cdots)$
$$= 70 \cdot 69$$
$$= 4830$$

NOTE: This technique of "cancelling out" or "reducing" the problem by removing the factors $68 \cdot 67 \cdot 66 \cdots 1$ from both the numerator and the denominator should be preferred over actually evaluating 70!, actually evaluating 68!, and then dividing those two very large numbers. In general, a smaller factorial in the denominator can be completely divided into a larger factorial in the numerator to leave only the "excess" factors not appearing the in the denominator. This is the technique employed in this manual -- e.g., see #9 below, where the 7! is cancelled from both the numerator and the denominator. In addition, the, the answer to a <u>counting</u> problem (but not a <u>probability</u> problem) must always be a whole number; a fractional number indicates that a mistake has been made.

5. $(9-3)! = (6)!$
$$= 6 \cdot 5 \cdot 4 \cdot 3 \cdot 2 \cdot 1$$
$$= 720$$

7. $_6P_2 = 6!/(6-2)!$
$$= 6!/4!$$
$$= 6 \cdot 5$$
$$= 30$$

9. $_{10}C_3 = 10!/(7!3!)$
$$= (10 \cdot 9 \cdot 8)/3! \qquad \text{[see the NOTE in \#3 above]}$$
$$= 720/6$$
$$= 120$$

11. $_{52}C_2$ = 52!/(50!2!)
    = (52·51)/2!   [see the NOTE in #3 above]
    = 2652/2
    = 1326

13. $_nP_n$ = n!/(n-n)!
    = n!/0!
    = n!/1
    = n!

15. $_nC_0$ = n!/(n!0!)
    = n!/(n!·1)
    = n!/n!
    = 1

    NOTE: In words, this represents the number of ways to choose zero objects from among n objects. There is only 1 way -- to leave them all unselected.

17. (3)(2) = 6

19. $_{10}P_4$ = 10!/(10-4)!
    = 10!/6!
    = 10·9·8·7
    = 5040

21. $_{22}C_{12}$ = 22!/(10!12!)
    = (22·21·20·19·18·17·16·15·14·13)/10!
    = 646,646

23. 10! = 10·9·8·7·6·5·4·3·2·1
    = 3,628,800

25. a. 7! = 5040
    b. Since only 1 of the 5040 possibilities can be the correct alphabetic order, the probability of that arrangement occurring is 1/5040.

27. Because each of the 9 digits in the SS number could be any one of the 10 possibilities 0,1,2,3,4,5,6,7,8,9, there are $10·10·10·10·10·10·10·10·10 = 10^9 = 1,000,000,000$ (i.e., one billion) possible SS numbers.

29. $_{16}P_4$ = 16!/(16-4)!
    = 16!/12!
    = 16·15·14·13
    = 43,680

31. There are $50·50·50 = 50^3 = 125,000$ sequences possible. While such sequences are popularly called "combinations," they are really (i.e., in correct mathematical terminology) permutations.

33. The selections within each category can be thought of as a combination problem (i.e., choosing so many from a given group when the order is not important). The final menu can then be thought of as a sequence of four category selections to which the fundamental counting rule may be applied.
    $_{10}C_1 · _8C_1 · _{13}C_2 · _3C_1$ = 10·8·78·3
    = 18,720

35. a. Assuming the order of the shows is irrelevant in this context, use combinations.

$$_{14}C_5 = 14!/(9!5!)$$
$$= (14 \cdot 13 \cdot 12 \cdot 11 \cdot 10)/5!$$
$$= 2002$$

   b. There are 2002 - 650 = 1352 compatible combinations; the probability of selecting a compatible combination by random selection is therefore 1352/2002 = .675.

37. The number of possible combinations is $_{49}C_6 = 49!/(43!/6!)$
$$= (49 \cdot 48 \cdot 47 \cdot 46 \cdot 45 \cdot 44)/6!$$
$$= 13,983,816$$

The probability of winning with one random selection is therefore 1/13,983,816 -- about 1/20 of the 1/700,000 probability of getting hit with lightning during the year. This means a person is 20 times more likely to be hit by lightning during the year than to win such a lottery. NOTE: This does not necessarily mean that there should be about 20 times as many people in the state that are hit with lightning than that win the lottery. The number of lottery winners depends on the number that actually play, not only on the probability of winning. If few people play the lottery, there are few winners; if many people play, there are many winners -- we expect about 1 winner for every 13,983,816 tickets purchased. The probability of winning the lottery with one ticket, the probability of being hit by lightning in one year, and the number of people hit by lightning in one year are not so subject to human manipulation.

39. $24! = 24 \cdot 23 \cdot 22 \cdots 3 \cdot 2 \cdot 1$
$$= 6.20 \times 10^{23}$$

41. $10!/(7!3!) = (10 \cdot 9 \cdot 8)/3!$
$$= 120$$
NOTE: This is the same as $_{10}C_3$, the number of combinations of 3 objects chosen from among 10 objects. Picking a committee of 3 from 10 people, for example, can be thought of lining up the 10 people (e.g, in alphabetical order) and marking 3 as successes (committee members) and 7 as failures (nonmembers). Each different arrangement of successes and failures corresponds to a particular committee of 3.

43. The numbers of possible names with 1,2,3,...,8 letters must be calculated separately using the fundamental counting rule and then added together to determine the total number of possible names. A chart will help to organize the work.

# of letters	# of possible names		
1	26	= 26 =	26
2	$26 \cdot 36$	$= 26 \cdot 36^1 =$	936
3	$26 \cdot 36 \cdot 36$	$= 26 \cdot 36^2 =$	33,696
4	$26 \cdot 36 \cdot 36 \cdot 36$	$= 26 \cdot 36^3 =$	1,213,056
5	$26 \cdot 36 \cdots 36$	$= 26 \cdot 36^4 =$	43,670,016
6	$26 \cdot 36 \cdots 36$	$= 26 \cdot 36^5 =$	1,572,120,576
7	$26 \cdot 36 \cdots 36$	$= 26 \cdot 36^6 =$	56,596,340,736
8	$26 \cdot 36 \cdots 36$	$= 26 \cdot 36^7 =$	2,037,468,266,496
			2,095,681,645,538

45. Visualize the people entering one at a time, each person sitting to the right of the person who entered before him. Where the first person sits is irrelevant -- i.e., it merely establishes a reference point but does not affect the number of seating arrangements.

   a. Once the first person sits, there are 2 possibilities for the person at his right. Once the second person sits, there is only one possibility for the person at his right. Therefore, there are $2 \cdot 1 = 2$ possible arrangements.

b. Once the first person sits, there are n-1 possibilities for the person at his right. Once the second person sits, there are n-2 possibilities for the person at his right...[and so on]. Once the next to last person sits, there is only one possibility for the person at his right. Therefore, there are $(n-1) \cdot (n-2) \cdots 1 = (n-1)!$ possible arrangements.

47. The calculator factorial key gives $50! = 3.04140932 \times 10^{64}$.
Using the approximation, $K = (50.5) \cdot \log(50) + .39908993 - .43429448(50)$
$$= 85.79798522 + .39930993 - 21.71472400$$
$$= 64.48257115$$
And then $n! = 10^K$
$$= 10^{64.48257115}$$
$$= 3.037883739 \times 10^{64}$$
NOTE: The two answers differ by 3.5×10^{61} (i.e., by 35 followed by 60 zeros -- "zillions and zillions"). While that error may seem large, the numbers being dealt with are very large and that difference is only a $(3.5 \times 10^{61})/(3.04 \times 10^{64}) = 1.2\%$ error.

49. a. Assuming the judge knows there are to be 4 computers and 4 humans and frames his guess accordingly, the number of possible ways he could guess is the number of ways 4 of the 8 could be labeled as "computer" and is given by $_8C_4 = 8!/(4!4!)$
$$= (8 \cdot 7 \cdot 6 \cdot 5)/4!$$
$$= 70$$
The probability he guesses the right combination by chance alone is therefore 1/70.
NOTE: If the judge merely guesses on each one individually, either not knowing or not considering that there should be 4 computers and 4 humans, then the probability he guesses all 8 correctly is $(1/2)^8 = 1/256$.
b. Under the assumptions of part a, the probability that all 10 judges make all correct guesses is $(1/70)^8 = 3.54 \times 10^{-19}$.
NOTE: In the statement of the problem, the author states the Turing criterion to be "the person believes he or she is communicating with another person instead of a computer." In the comment in parentheses in part (b), he appears to take the criterion to be "the person cannot tell whether he or she is communicating with another person or with a computer." These are not the same criteria. Part of the problem is the ambiguous directive that the judges "cannot distinguish between computers and people" -- which could mean either "the judges recognize people, but also think the computers are people" or "the judges aren't sure whether any one of the communicators is a person or a computer." This, coupled with the NOTE in part (a), indicates some of the complications that often arise in real research and the necessity for precision in terminology before attempting any statistical analysis.

Review Exercises

1. let M = the driver is a male
a. P(all males) $= P(M_1$ and M_2 and M_3 and M_4 and $M_5)$
$$= P(M_1) \cdot P(M_2) \cdot P(M_3) \cdot P(M_4) \cdot P(M_5)$$
$$= (.90)(.90)(.90)(.90)(.90)$$
$$= .590$$
b. P(at least one female) $= P(\overline{\text{all males}})$
$$= 1 - P(\text{all males})$$
$$= 1 - .590$$
$$= .410$$

3. let S = a hang glider survives for the year
$$P(S) = 1 - .008$$
$$= .992$$
  a. P(all survive) = $P(S_1$ and S_2 and...and $S_{10})$
$$= P(S_1) \cdot P(S_2) \cdots P(S_{10})$$
$$= (.992)(.992)...(.992)$$
$$= (.992)^{10}$$
$$= .923$$
  b. P(at least does not survive) = P($\overline{\text{all survive}}$)
$$= 1 - P(\text{all survive})$$
$$= 1 - .923$$
$$= .077$$

5. It may be helpful to make a chart like the one given at the right.

let F = a person is female
   Y = a person uses seat belts

		SEAT BELTS?		
		Yes	No	
SEX	Male	29	21	50
	Female	69	31	100
		98	52	150

P(F or Y) = P(F) + P(Y) - P(F and Y)
$$= 100/150 + 98/150 - 69/150$$
$$= 129/150$$
$$= .86$$

7. Since 2945 of the 6665 films have an R rating, the probability that a single randomly selected film is one with an R rating is 2945/6665 = .442.

9. The number of different possible combinations of 6 such numbers is
$$_{40}C_6 = 40!/(34!6!)$$
$$= (40 \cdot 39 \cdot 38 \cdot 37 \cdot 36 \cdot 35)/6!$$
$$= 3,838,380$$
The probability of winning with one random selection is therefore 1/3,838,380.

11. let D = the selected circuit is defective
  a. $P(D_1$ and $D_2) = P(D_1) \cdot P(D_2 \mid D_1)$
$$= (18/120) \cdot (18/120)$$
$$= .0225$$
NOTE: The "and" formula used above <u>always</u> works. When the events are independent, $P(D_2 \mid D_1) = P(D_2) = P(D_1) = P(D)$ is always 18/120 and the formula for independent events (really a special case of the above formula) can be used.
$$P(D_1 \text{ and } D_2) = P(D_1) \cdot P(D_2)$$
$$= (18/120) \cdot (18/120)$$
$$= .0225$$
  b. $P(D_1$ and $D_2) = P(D_1) \cdot P(D_2 \mid D_1)$
$$= (18/120) \cdot (17/119)$$
$$= .0214$$

13. let R = a Republican is selected

$P(R) = .30$ for each selection

$P(R_1 \text{ and } R_2 \text{ and...and } R_{12}) = P(R_1) \cdot P(R_2) \cdot \cdot \cdot P(R_{12})$
$$= (.30) \cdot (.30) \cdot \cdot \cdot (.30)$$
$$= (.30)^{12}$$
$$= .000000531$$

NOTE: $P(R) = .30$ for each selection assumes independence, which technically is not so. If selecting a voter means that voter cannot be selected again during subsequent selections,

$P(R_1) = 60,000/200,000 = .3000000$

$P(R_2) = 59,999/199,999 = .2999965$

$P(R_3) = 59,998/199,998 = .2999930$

and so on

The guideline in the text suggests ignoring the lack of independence whenever the total sample is less than 5% of the population. In this case, the sample is $12/200,000 = .006\%$ of the population, well within the suggested limits for proceeding under the assumption of independence.

15. a. for independent events, $P(A \text{ and } B) = P(A) \cdot P(B)$
$$= (.2) \cdot (.4)$$
$$= .08$$

b. for mutually exclusive events, $P(A \text{ or } B) = P(A) + P(B)$
$$= .2 + .4$$
$$= .6$$

c. in all cases, $P(\bar{A}) = 1 - P(A)$
$$= 1 - .2$$
$$= .8$$

17. There are $5! = 5 \cdot 4 \cdot 3 \cdot 2 \cdot 1 = 120$ possible orderings, only one of which is correct. The probability of obtaining the correct ordering by a single random guess, therefore, is 1/120.

19. let F = the applicant selected is female

    M = the applicant is male

$P(F) = P(M) = .5$, for each selection

a. $P(\text{all women}) = P(F_1 \text{ and } F_2 \text{ and } F_3 \text{ and } F_4)$
$$= P(F_1) \cdot P(F_2) \cdot P(F_3) \cdot P(F_4)$$
$$= (.5) \cdot (.5) \cdot (.5) \cdot (.5)$$
$$= .0625$$

b. $P(\text{all men}) = P(M_1 \text{ and } M_2 \text{ and } M_3 \text{ and } M_4)$
$$= P(M_1) \cdot P(M_2) \cdot P(M_3) \cdot P(M_4)$$
$$= (.5) \cdot (.5) \cdot (.5) \cdot (.5)$$
$$= .0625$$

c. $P(\text{same sex}) = P(\text{all women or all men})$ [NOTE: these events are mutually exclusive]
$$= P(\text{all women}) + P(\text{all men})$$
$$= .0625 + .0625$$
$$= .1250$$

Chapter 4

Probability Distributions

4-2 Random Variables

1. This is a probability distribution since $\Sigma P(x)=1$ is true and $0 \leq P(x) \leq 1$ is true for each x.

x	P(x)	x·P(x)	x^2	x^2·P(x)
4	.120	.480	16	1.920
5	.253	1.265	25	6.325
6	.217	1.302	36	7.812
7	.410	2.870	49	20.090
	1.000	5.917		36.147

$\mu = \Sigma x \cdot P(x)$
$= 5.9$
$\sigma^2 = \Sigma x^2 \cdot P(x) - \mu^2$
$= 36.147 - (5.917)^2$
$= 1.136$ rounded to 1.1
$\sigma = 1.1$

NOTE: Several important statements should be kept in mind when working with probability distributions and the above formulas.
* If one of the conditions for a probability distribution does not hold, the formulas do not apply and produce numbers that have no meaning.
* $\Sigma x \cdot P(x)$ gives the mean of the x values and must be a number between the highest and lowest x values.
* $\Sigma x^2 \cdot P(x)$ gives the mean of the x^2 values and must be a number between the highest and lowest x^2 values.
* $\Sigma P(x)$ must always equal 1.000.
* Σx and Σx^2 have no meaning and should not be calculated.
* The quantity $[\Sigma x^2 \cdot P(x) - \mu^2]$ cannot possibly be negative; if it is, then there is a mistake.
* Always be careful to use the <u>unrounded</u> mean in the calculation of the variance and to take the square root of the <u>unrounded</u> variance to find the standard deviation.

3. This is <u>not</u> a probability distribution since $\Sigma P(x) = .78 \neq 1.00$.

5. This is a probability distribution since $\Sigma P(x)=1$ is true and $0 \leq P(x) \leq 1$ is true for each x.

x	P(x)	x·P(x)	x^2	x^2·P(x)
0	.125	0	0	0
1	.375	.375	1	.375
2	.375	.750	4	1.500
3	.125	.375	9	1.125
	1.000	1.500		3.000

$\mu = \Sigma x \cdot P(x)$
$= 1.5$
$\sigma^2 = \Sigma x^2 \cdot P(x) - \mu^2$
$= 3.000 - (1.500)^2$
$= .75$
$\sigma = .9$

7. This is a probability distribution since $\Sigma P(x)=1$ is true and $0 \leq P(x) \leq 1$ is true for each x.

x	P(x)	x·P(x)	x^2	x^2·P(x)
0	.26	0	0	0
1	.16	.16	1	.16
2	.12	.24	4	.48
3	.09	.27	9	.81
4	.07	.28	16	1.12
5	.09	.45	25	2.25
6	.07	.42	36	2.52
7	.14	.98	49	6.86
	1.00	2.80		14.20

$\mu = \Sigma x \cdot P(x)$
  $= 2.8$
$\sigma^2 = \Sigma x^2 \cdot P(x) - \mu^2$
  $= 14.20 - (2.80)^2$
  $= 6.36$, rounded to 6.4
$\sigma = 2.5$

9. This is a probability distribution since $\Sigma P(x)=1$ is true and $0 \leq P(x) \leq 1$ is true for each x.

x	P(x)	x·P(x)	x^2	x^2·P(x)
0	.36	0	0	0
1	.48	.48	1	.48
2	.16	.32	4	.64
	1.00	.80		1.12

$\mu = \Sigma x \cdot P(x)$
  $= .8$
$\sigma^2 = \Sigma x^2 \cdot P(x) - \mu^2$
  $= 1.12 - (.80)^2$
  $= .48$, rounded to .5
$\sigma = .7$

11. This is a probability distribution since $\Sigma P(x)=1$ is true and $0 \leq P(x) \leq 1$ is true for each x.

x	P(x)	x·P(x)	x^2	x^2·P(x)
0	.026	0	0	0
1	.345	.345	1	.345
2	.346	.692	4	1.384
3	.154	.462	9	1.386
4	.129	.516	16	2.064
	1.000	2.011		5.179

$\mu = \Sigma x \cdot P(x)$
  $= 2.0$
$\sigma^2 = \Sigma x^2 \cdot P(x) - \mu^2$
  $= 5.179 - (2.011)^2$
  $= 1.139$, rounded to 1.1
$\sigma = 1.1$

13.

x	P(x)	x·P(x)
500	.25	125
200	.75	150
	1.00	275

$E = \Sigma x \cdot P(x)$
$= \$275$

15.

x	P(x)	x·P(x)	E = Σx·P(x)
-30	.15	-4.50	= $97.50
120	.85	102.00	
	1.00	97.50	

NOTE: Giving precedence to the standard notation for reporting monetary values over the rounding guideline stated in the text, we give two decimal place accuracy in the answer.

17. This problem can be worked in two different ways.

* Considering the $156 she paid up front, the woman loses 156 if she lives and gains 100,000 - 156 = 99844 if she dies.

x	P(x)	x·P(x)	E = Σx·P(x)
-156	.9995	-155.92	= -$106.00 (i.e., a loss of $106)
99844	.0005	44.92	
	1.0000	-106.00	

* Ignoring the cost of the insurance, one may calculate the expected value of the policy itself.

x	P(x)	x·P(x)	E = Σx·P(x)
0	.9995	0	= $50.00
100000	.0005	50.00	
	1.0000	50.00	

If the policy is worth $50 in expected returns and the woman pays $156 to purchase it, her net expectation is 50 - 156 = -$106.

19. The 8 equally like outcomes in the sample space are given at the right. If x represents the number of girls, counting the numbers of favorable outcomes indicates

$P(x = 0) = 1/8 = .125$
$P(x = 1) = 3/8 = .375$
$P(x = 2) = 3/8 = .375$
$P(x = 3) = 1/8 = .125$

outcome	x
BBB	0
BBG	1
BGB	1
GBB	1
GGB	2
GBG	2
BGG	2
GGG	3

x	P(x)	x·P(x)	x²	x²·P(x)
0	.125	0	0	0
1	.375	.375	1	.375
2	.375	.750	4	1.500
3	.125	.375	9	1.125
	1.000	1.500		3.000

$\mu = \Sigma x \cdot P(x)$
$\quad = 1.5$
$\sigma^2 = \Sigma x^2 \cdot P(x) - \mu^2$
$\quad = 3.000 - (1.500)^2$
$\quad = .75$
$\sigma = .9$

21. There are 4 possible outcomes, but they are not equally likely. If G is a good (i.e., "made") shot, B is a bad (i.e., "missed") shot, and x represents the number of shots made, then
$P(G) = .85$
$P(B) = .15$
and the following summary table applies.

outcome	probability					x
GG	$P(G_1$ and $G_2)$	$= P(G_1) \cdot P(G_2)$	$= (.85)(.85)$	$=$	.7225	2
GB	$P(G_1$ and $B_2)$	$= P(G_1) \cdot B(G_2)$	$= (.85)(.15)$	$=$	.1275	1
BG	$P(B_1$ and $G_2)$	$= P(B_1) \cdot P(G_2)$	$= (.15)(.85)$	$=$	.1275	1
BB	$P(B_1$ and $B_2)$	$= P(B_1) \cdot P(B_2)$	$= (.15)(.15)$	$=$	.0225	0
					1.0000	

x	P(x)	$x \cdot P(x)$	x^2	$x^2 \cdot P(x)$
0	.0225	0	0	0
1	.2550	.2550	1	.2550
2	.7225	1.4450	4	2.8900
	1.0000	1.7000		3.1450

$\mu = \Sigma x \cdot P(x)$
$= 1.7$
$\sigma^2 = \Sigma x^2 \cdot P(x) - \mu^2$
$= 3.1450 - (1.7000)^2$
$= .2550$, rounded to .3
$\sigma = .5$

23. In order for P(x) to be a probability distribution, it must be true that
(1) $\Sigma P(x) = 1$
(2) $0 \leq P(x) \leq 1$ for every permissible value of x.
NOTE: Parts (a) and (b) below use the fact from algebra that
$1 + r + r^2 + r^3 + r^4 + ... = 1/(1-r)$ for any r such that $-1 < r < 1$

a. P(x) is a probability distribution because for x = 1,2,3...
(1) $\Sigma P(x) = \Sigma(.4)(.6)^{x-1}$
$= (.4) \cdot \Sigma(.6)^{x-1}$
$= (.4) \cdot [1 + .6 + .6^2 + .6^3 + ...]$
$= (.4) \cdot [1/(1-.6)]$
$= (.4) \cdot [1/.4]$
$= 1$
(2) $0 \leq P(x) \leq 1$, since P(x) > 0 for all x
and P(x) = .4 for x = 1
and P(x) keeps decreasing as x grows larger

b. P(x) is a probability distribution because for x = 1,2,3...
(1) $\Sigma P(x) = \Sigma(.5)^x$
$= (.5) \cdot \Sigma(.5)^{x-1}$
$= (.5) \cdot [1 + .5 + .5^2 + .5^3 + ...]$
$= (.5) \cdot [1/(1-.5)]$
$= (.5) \cdot [1/.5]$
$= 1$
(2) $0 \leq P(x) \leq 1$, since P(x) > 0 for all x
and P(x) = .5 for x = 1
and P(x) keeps decreasing as x grows larger

c. P(x) is not a probability distribution since for x = 1,2,3,...
   (1) $\Sigma P(x) = \Sigma[1/(2x)]$
       $= 1/2 + 1/4 + 1/6 + 1/8 + ...$
       $= .5000 + .2500 + .1666 + .1250 + ...$
       $= 1.0416 + ...$
       > 1
   (2) [Condition (2) happens to be satisfied, but that is irrelevant since condition (1) is not satisfied and both conditions must be met to have a probability distribution.]

25. a. Exercise 21(c) of section 2-5 notes that "multiplying each score by a constant also multiplies the distances between scores by that constant, and hence the measures of dispersion in the original units are multiplies by that constant (and the variance is multiplied by the square of that constant)." Multiplying each score by 5, therefore, multiplies the variance by $5^2 = 25$ and the new variance is $25 \cdot (1.25) = 31.25$.
    b. Refer to part (a). Dividing by 5 is the same as multiplying by 1/5. Multiplying each score by 1/5, therefore, multiplies the variance by $(1/5)^2 = 1/25 = .04$ and the new variance is $.04 \cdot (1.25) = .05$.
    c. Exercise 21(a) of section 2-5 notes that "adding a constant to each score does not affect the spread of the scores of the measures of dispersion." Adding 5 to each score, therefore, does not change the variance and it remains 1.25.
    d. Exercise 21(b) of section 2-5 notes that "subtracting a constant from each score does not affect the spread of the scores of the measures of dispersion." Subtracting 5 from each score, therefore, does not change the variance and it remains 1.25. NOTE: An alternative answer is "Refer to part (c). Subtracting 5 from each score is the same as adding -5. Adding -5 to each score, therefore, does not change the variance and it remains 1.25."

27. The equivalence between the two expressions may be demonstrated in 7 steps as follows.
    (1) $\Sigma(x-\mu)^2 \cdot P(x) = \Sigma(x^2 - 2\mu x + \mu^2) \cdot P(x)$
    (2) $\qquad\qquad = \Sigma(x^2 \cdot P(x) - 2\mu x \cdot P(x) + \mu^2 \cdot P(x))$
    (3) $\qquad\qquad = \Sigma x^2 \cdot P(x) - \Sigma 2\mu x \cdot P(x) + \Sigma \mu^2 \cdot P(x)$
    (4) $\qquad\qquad = \Sigma x^2 \cdot P(x) - 2\mu \cdot \Sigma x \cdot P(x) + \mu^2 \cdot \Sigma P(x)$
    (5) $\qquad\qquad = \Sigma x^2 \cdot P(x) - 2\mu \cdot \mu + \mu^2 \cdot \Sigma P(x)$
    (6) $\qquad\qquad = \Sigma x^2 \cdot P(x) - 2\mu^2 + \mu^2$
    (7) $\qquad\qquad = \Sigma x^2 \cdot P(x) - \mu^2$

The algebraic justification for each of the preceding steps is as follows.
(1) $(a-b)^2 = a^2 - 2ab + b^2$
(2) $(a-b-c)d = ad - bd - cd$
(3) $\Sigma(A - B + C) = \Sigma A - \Sigma B + \Sigma C$
(4) $\Sigma cX = c\Sigma X$, for any constant c [and μ is a constant]
(5) $\Sigma x \cdot P(x) = \mu$
(6) $\Sigma P(x) = 1$, for any probability distribution
(7) $-2a + a = -a$

4-3 Binomial Experiments

NOTE: To use the binomial formula, one must identify 3 quantities: n,x,p. Table A-1, for example, requires only these 3 values to supply a probability. Since what the text calls "q" always equals 1-p, it can be so designated without introducing unnecessary notation [just as no special notation is utilized for the quantity n-x, even though it appears twice in the binomial formula]. Accordingly, in the interest of simplicity, this manual uses the formula

$$P(x) = [n!/x!(n-x)!] \cdot p^x \cdot (1-p)^{n-x}$$

without introducing a special name for the quantity 1-p. This has the additional advantage of ensuring that the probabilities p and 1-p sum to 1.00 and protecting against an error in the separate calculation and/or identification of "q." Interestingly, this 1-p notation is employed in the text when presenting the geometric distribution in exercise #33 of this section.

1. The 4 requirements are:
   #1 There are a fixed number of trials.
   #2 The trials are independent.
   #3 Each trial has two possible named outcomes.
   #4 The probabilities remain constant for each trial.
   a. Yes, all 4 requirements are met.
   b. No, requirements #2 and #4 are not met. The probability the first lens is defective is 8/20. The probability the second lens is defective is either 8/19 or 7/19, depending on the first selection.
   c. Yes, all 4 requirements are met.
   d. Yes, all 4 requirements are met.
   e. Yes, all 4 requirements are met. NOTE: Assuming the selection process and/or the actual conducting of the experiment would not all ow the same person to be selected twice, the scenario is very technically not a binomial experiment. Assuming the population under consideration is so large that removing the people already sampled from the population does not appreciably affect the population or the probabilities, however, the binomial formulas may be applied.

3. From table A-1, .101.

5. From table A-1, .349.

7. $P(x) = [n!/x!(n-x)!] \cdot p^x \cdot (1-p)^{n-x}$
   $= [5!/3!2!] \cdot (1/4)^3 \cdot (3/4)^2$
   $= [10] \cdot (1/64) \cdot (9/16)$
   $= 90/1024$ or $45/512$

9. $P(x) = [n!/x!(n-x)!] \cdot p^x \cdot (1-p)^{n-x}$
   $= [6!/2!4!] \cdot (1/2)^2 \cdot (1/2)^4$
   $= [15] \cdot (1/4) \cdot (1/16)$
   $= 15/64$

11. let x = the number of girls
    n = 10
    x = 4
    p = .5
    $P(x) = [n!/x!(n-x)!] \cdot p^x \cdot (1-p)^{n-x}$
    $= [10!/4!6!] \cdot (.5)^4 \cdot (.5)^6$
    $= [210] \cdot (.0625) \cdot (.015625)$
    $= .205$

IMPORTANT NOTE: The intermediate values of 210, .0625 and .015625 are given above to help students with an incorrect answer to identify the portion of the problem in which the mistake was made. This practice will be followed in all problems (i.e., not just binomial problems) throughout the manual. In practice, all calculations can be done in one step on the calculator. You may choose to (or be asked to) write down such intermediate values for your own (or the instructor's) benefit, but <u>never round off in the middle of a problem.</u> <u>Do not write the values down on paper and then re-enter them in the calculator -- use the memory to let the calculator remember with complete accuracy any intermediate values that will be used in subsequent calculations.</u> In addition, always make certain that the quantity [n!/x!(n-x)!] is a whole number and that the final answer is between 0 and 1.

13. n = 3
    x = 2
    p = .01
    $P(x) = [n!/x!(n-x)!] \cdot p^x \cdot (1-p)^{n-x}$
    $= [3!/2!1!] \cdot (.01)^2 \cdot (.99)^1$
    $= [3] \cdot (.0001) \cdot (.99)$
    $= .000297$

15. let x = the number that are late
    n = 10
    x = 2 or more
    p = .10
    NOTE: Since x represents the number that are late (i.e., being late is considered a "success"), then p must represent the probability of being late. Beginning each problem with a statement identifying what x represents helps to avoid errors caused by inconsistent identifications. The probabilities below come from Table A-1.
    $P(x \geq 2) = 1 - P(x \leq 1)$
    $= 1 - [P(x=0) + P(x=1)]$
    $= 1 - [.349 + .387]$
    $= 1 - .736$
    $= .264$

17. let x = the number that arrive for the flight
    n = 14
    x = 13 or more
    p = .90
    $P(x \geq 13) = P(x=13) + P(x=14)$
    $= .356 + .229$ [from Table A-1]
    $= .585$

19. let x = the number that survive
   n = 20
   x = 18
   p = .95
   $P(x) = [n!/x!(n-x)!] \cdot p^x \cdot (1-p)^{n-x}$
   $= [20!/18!2!] \cdot (.95)^{18} \cdot (.05)^2$
   $= [190] \cdot (.3972) \cdot (.0025)$
   $= .189$

21. let x = the number with HS diploma but no college
   n = 15
   x = at least 10
   p = .40
   $P(x \geq 10) = P(x=10) + P(x=11) + P(x=12) + P(x=13) + P(x=14) + P(x=15)$
   $= .024 + .007 + .002 + .000^+ + .000^+ + .000^+$ [from Table A-1]
   $= .033$

23. let x = the number living within 50 miles of a coastal shoreline
   n = 20
   x = 12
   p = .53
   $P(x) = [n!/x!(n-x)!] \cdot p^x \cdot (1-p)^{n-x}$
   $= [20!/12!8!] \cdot (.53)^{12} \cdot (.47)^8$
   $= [125970] \cdot (.000491) \cdot (.00238)$
   $= .147$

25. a. let x = the number that contain errors
      n = 50
      x = 50
      p = .98
      $P(x) = [n!/x!(n-x)!] \cdot p^x \cdot (1-p)^{n-x}$
      $= [50!/50!0!] \cdot (.98)^{50} \cdot (.02)^0$
      $= [1] \cdot (.3642) \cdot (1.0000)$
      $= .364$
   b. n = 50
      x = 49
      p = .98
      $P(x) = [n!/x!(n-x)!] \cdot p^x \cdot (1-p)^{n-x}$
      $= [50!/49!1!] \cdot (.98)^{49} \cdot (.02)^1$
      $= [50] \cdot (.3716) \cdot (.0200)$
      $= .372$

27. a. let x = the number that arrive on time
      n = 12
      x = at least 9
      p = .60
      $P(x \geq 9) = P(x=9) + P(x=10) + P(x=11) + P(x=12)$
      $= .142 + .064 + .017 + .002$ [from Table A-1]
      $= .225$

b. let x = the number that arrive on time
   n = 30
   x = 20
   p = .60
   $P(x) = [n!/x!(n-x)!] \cdot p^x \cdot (1-p)^{n-x}$
   $= [30!/20!10!] \cdot (.60)^{20} \cdot (.5)^{10}$
   $= [30,045,015] \cdot (.0000366) \cdot (.000105)$
   $= .115$

NOTE: In mathematics and statistics, including the Poisson formula, the letter e is commonly used for Euler's constant and has the value (accurate to five decimal places) e = 2.71828. Most calculators have an "e^x" key to find e raised to a power -- or one could also use the y^x key to raise 2.71828 to the desired power. Since any non-zero constant raised to the zero power is 1, $e^0 = 1$. In the Poisson formula, e will always be raised to a non-positive power and $e^{-\mu}$ will always be between 0 and 1.

29. let x = the number of births per day
    $\mu = 2.25$
    a. x = 0
       $P(x) = \mu^x \cdot e^{-\mu}/x!$
       $= (2.25)^0 \cdot e^{-2.25}/0!$
       $= (1) \cdot (.1054)/1$
       $= .105$
    b. x = 1
       $P(x) = \mu^x \cdot e^{-\mu}/x!$
       $= (2.25)^1 \cdot e^{-2.25}/1!$
       $= (2.25) \cdot (.1054)/1$
       $= .237$
    c. x = 4
       $P(x) = \mu^x \cdot e^{-\mu}/x!$
       $= (2.25)^4 \cdot e^{-2.25}/4!$
       $= (25.629) \cdot (.1054)/24$
       $= .113$

31. let x = the number of defects
    $\mu = 2.0$
    x = more than 1
    $P(x \geq 1) = 1 - P(x \leq 1)$
    $= 1 - [P(x=0) + P(x=1)]$
    $= 1 - [\mu^0 \cdot e^{-\mu}/0! + \mu^1 \cdot e^{-\mu}/1!]$
    $= 1 - [(2.0)^0 \cdot e^{-2.0}/0! + (2.0)^1 \cdot e^{-2.0}/1!]$
    $= 1 - [(1) \cdot (.1353)/1 + (2.0) \cdot (.1353)/1]$
    $= 1 - [.1353 + .2707]$
    $= 1 - .4060$
    $= .594$

33. let x = the number of components tested to find 1st defect
    x = 7
    p = .2
    $P(x) = p \cdot (1-p)^{x-1}$
    $= (.2) \cdot (.8)^6$
    $= (.2) \cdot (.2621)$
    $= .0524$

35. a. Extending the pattern to cover 6 types of outcomes, where $\Sigma x = n$ and $\Sigma p = 1$,

$$P(x_1,x_2,x_3,x_4,x_5,x_6) = [n!/(x_1!x_2!x_3!x_4!x_5!x_6!)] \cdot p_1^{x_1} \cdot p_2^{x_2} \cdot p_3^{x_3} \cdot p_4^{x_4} \cdot p_5^{x_5} \cdot p_6^{x_6}$$

b. $n = 20$

$x_1 = 5$, $x_2 = 4$, $x_3 = 3$, $x_4 = 2$, $x_5 = 3$, $x_6 = 3$

$p_1 = p_2 = p_3 = p_4 = p_5 = p_6 = 1/6$

$$\begin{aligned}
P(x_1,x_2,x_3,x_4,x_5,x_6) &= [n!/(x_1!x_2!x_3!x_4!x_5!x_6!)] \cdot p_1^{x_1} \cdot p_2^{x_2} \cdot p_3^{x_3} \cdot p_4^{x_4} \cdot p_5^{x_5} \cdot p_6^{x_6} \\
&= [20!/(5!4!3!2!3!3!)] \cdot (1/6)^5 \cdot (1/6)^4 \cdot (1/6)^3 \cdot (1/6)^2 \cdot (1/6)^3 \cdot (1/6)^3 \\
&= [20!/(5!4!3!2!3!3!)] \cdot (1/6)^{20} \\
&= [1.955 \cdot 10^{12}] \cdot (2.735 \cdot 10^{-16}) \\
&= .000535
\end{aligned}$$

37. a. binomial

let x = the number of wins in 20 spins

$n = 20$

$x = 1$

$p = 1/38$

$$\begin{aligned}
P(x) &= [n!/x!(n-x)!] \cdot p^x \cdot (1-p)^{n-x} \\
&= [20!/1!19!] \cdot (1/38)^1 \cdot (19/38)^{19} \\
&= [20] \cdot (.0263) \cdot (.6025) \\
&= .317
\end{aligned}$$

b. Poisson

let x = the number of wins in 20 spins

$\mu = 20/28$

NOTE: If the probability of a win on any one spin is 1/38, then the expected number of wins is n/38 for n spins (i.e., 1/38 for 1 spin, 38/38 = 1 for 38 spins, 20/38 for 30 spins, etc.). Keep in mind that this is the <u>expected number</u> of wins in n spins, not the <u>probability</u> of winning in n spins.

$x = 1$

$$\begin{aligned}
P(x) &= \mu^x \cdot e^{-\mu}/x! \\
&= (20/30)^1 \cdot e^{-20/38}/1! \\
&= (.5263)^1 \cdot e^{-.5263}/1! \\
&= (.5263) \cdot (.5908)/1 \\
&= .311
\end{aligned}$$

4.4 Mean, Variance and Standard Deviation for the Binomial Distribution

1. $\mu = n \cdot p$

   $= (49) \cdot (.5)$

   $= 24.5$

   $\sigma^2 = n \cdot p \cdot (1-p)$

   $= (49) \cdot (.5) \cdot (.5)$

   $= 12.25$

   $\sigma = 3.5$

3. $\mu = n \cdot p$

   $= (473) \cdot (.855)$

   $= 404.4$

   $\sigma^2 = n \cdot p \cdot (1-p)$

   $= (473) \cdot (.855) \cdot (.145)$

   $= 58.6$

   $\sigma = 7.7$

5. let x = the number of correct answers
   n = 50
   p = .5
   $\mu = n \cdot p$
   $= (50) \cdot (.5)$
   $= 25.0$
   $\sigma^2 = n \cdot p \cdot (1-p)$
   $= (50) \cdot (.5) \cdot (.5)$
   $= 12.5$
   $\sigma = 3.5$

7. let x = the number of R-rated films
   n = 20
   p = 2945/6665 = .4419
   NOTE: Technically, the binomial formulas do not apply. Once a film is selected and removed from the population from which further selections are made, the probability of selecting an R-rated film is no longer 2945/6665. Since removing a sample of 20 from such a large population does not appreciably change the population or the probabilities, however, the binomial formulas may still be applied.
   $\mu = n \cdot p$
   $= (20) \cdot (.4419)$
   $= 8.8$
   $\sigma^2 = n \cdot p \cdot (1-p)$
   $= (20) \cdot (.4419) \cdot (.5581)$
   $= 4.9$
   $\sigma = 2.2$

9. let x = the number of injuries caused by seat failure
   n = 200
   p = .47
   $\mu = n \cdot p$
   $= (200) \cdot (.47)$
   $= 94.0$
   $\sigma^2 = n \cdot p \cdot (1-p)$
   $= (200) \cdot (.47) \cdot (.53)$
   $= 49.8$
   $\sigma = 7.1$

11. let x = the number of returns with taxpayer errors
    n = 12
    p = .37
    $\mu = n \cdot p$
    $= (12) \cdot (.37)$
    $= 4.4$
    $\sigma^2 = n \cdot p \cdot (1-p)$
    $= (12) \cdot (.37) \cdot (.63)$
    $= 2.8$
    $\sigma = 1.7$

13. a. let x = the number of pizzas delivered on time
     n = 300
     p = .90
     μ = n·p
        = (300)·(.90)
        = 270.0
     σ^2 = n·p·(1-p)
        = (300)·(.90)·(.10)
        = 27.0
     σ = 5.2

   b. Unusual values are those outside $\mu \pm 2 \cdot \sigma$
$$270.0 \pm 2 \cdot (5.2)$$
$$270.0 \pm 10.4$$
$$259.6 \text{ to } 280.4$$
   As 244 on-time deliveries is not within these limits, it would be considered an unusual event.

15. a. let x = the number of colors correctly guessed
     n = 50
     p = 1/6
     μ = n·p
        = (50)·(1/6)
        = 8.3
     σ^2 = n·p·(1-p)
        = (50)·(1/6)·(5/6)
        = 6.9
     σ = 2.6

   b. Unusual values are those outside $\mu \pm 2 \cdot \sigma$
$$8.3 \pm 2 \cdot (2.6)$$
$$8.3 \pm 5.2$$
$$3.1 \text{ to } 13.5$$
   As 12 correct guesses is within these limits, it would not be considered an unusual event.

17. let x = the number of accidents involving injuries
    n = 80
    p = .612
    μ = n·p
       = (80)·(.612)
       = 49.0
    σ^2 = n·p·(1-p)
       = (80)·(.612)·(.388)
       = 19.0
    σ = 4.4
   Unusual values are those outside $\mu \pm 2 \cdot \sigma$
$$49.0 \pm 2 \cdot (4.4)$$
$$49.0 \pm 8.8$$
$$40.2 \text{ to } 57.8$$
   As 72 accidents with injuries is not within these limits, it would be considered an unusual event.

19. a. let x = the number of consumers recognizing the Coke brand name

$n = 200$

$p = .95$

$\mu = n \cdot p$

$\quad = (200) \cdot (.95)$

$\quad = 191.0$

$\sigma^2 = n \cdot p \cdot (1-p)$

$\quad = (200) \cdot (.95) \cdot (.05)$

$\quad = 9.5$

$\sigma = 3.1$

b. Unusual values are those outside $\mu \pm 2 \cdot \sigma$

$$191.0 \pm 2 \cdot (3.1)$$
$$191.0 \pm 6.2$$
$$184.8 \text{ to } 197.2$$

As all 200 consumers recognizing the Coke brand name is not within these limits, it would be considered an unusual event.

21. a. let x = the number of correct responses

$n = 100$

$p = .20$

$\mu = n \cdot p$

$\quad = (100) \cdot (.20)$

$\quad = 20.0$

$\sigma^2 = n \cdot p \cdot (1-p)$

$\quad = (100) \cdot (.20) \cdot (.80)$

$\quad = 16.0$

$\sigma = 4.0$

b. $z_{28} = (x-\mu)/\sigma \qquad\qquad z_{12} = (x-\mu)/\sigma$

$\quad = (28-20)/4 \qquad\qquad\quad = (12-20)/4$

$\quad = 2.00 \qquad\qquad\qquad\quad = -2.00$

The z scores indicate that the range from 20 to 28 includes scores within 2 standard deviations of the mean. Since Chebyshev's Theorem (section 2-5) states that there must be at least $1 - 1/K^2$ of the scores within K standard deviations of the mean, there must be at least $1 - 1/2^2 = 1 - 1/4 = 3/4$ of the scores within $K=2$ standard deviations of the mean. In other words, we expect to find that at least 75% of the students who guess will score between 12 and 28. NOTE: Chebyshev's Theorem makes a statement about what must be true in a set of scores with a specific mean and standard deviation. While it can be used to make statements about what we may expect in a sample, it cannot make definitive predictions about samples using population and/or theoretical means and standard deviations.

c. $z_{30} = (x-\mu)/\sigma$

$\quad = (30-20)/4$

$\quad = 2.50$

Review Exercises

1. binomial problem
   let x = the number of students owning videocassette recorders
   n = 10
   p = .3
   a. x = 5
      P(x=5) = .103   [from Table A-1]
   b. x = at least 5
      $P(X \geq 5) = P(x=5) + P(x=6) + P(x=7) + P(x=8) + P(x=9) + P(x=10)$
      $= .103 + .037 + .009 + .001 + .000^+ + .000^+$ [from Table A-1]
      $= .150$
   c. $\mu = n \cdot p$
      $= (10) \cdot (.30)$
      $= 3.0$
      $\sigma^2 = n \cdot p \cdot (1-p)$
      $= (10) \cdot (.30) \cdot (.70)$
      $= 2.1$
      $\sigma = 1.4$

3. n = 10
   x = 6
   p = .05
   $P(x) = [n!/x!(n-x)!] \cdot p^x \cdot (1-p)^{n-x}$
   $= [10!/6!4!] \cdot (.05)^6 \cdot (.95)^4$
   $= [210] \cdot (.0000000156) \cdot (.8145)$
   $= .00000267$

5. binomial problem
   let x = the number of Florida passengers wearing seat belts

n = 5	n = 5
x = 4	x = 5
p = .57	p = .57
$P(x) = [n!/x!(n-x)!] \cdot p^x \cdot (1-p)^{n-x}$	$P(x) = [n!/x!(n-x)!] \cdot p^x \cdot (1-p)^{n-x}$
$= [5!/4!1!] \cdot (.57)^4 \cdot (.43)^1$	$= [5!/5!0!] \cdot (.57)^5 \cdot (.43)^0$
$= [5] \cdot (.1056) \cdot (.43)$	$= [1] \cdot (.0229) \cdot (1.00)$
$= .227$	$= .060$

x	P(x)	x·P(x)	x^2	$x^2 \cdot P(x)$
0	.015	0	0	0
1	.097	.097	1	.097
2	.258	.516	4	1.032
3	.342	1.026	9	3.078
4	.227	.908	16	3.632
5	.060	.300	25	1.500
	1.000	2.847		9.339

   $\mu = \Sigma x \cdot P(x)$
   $= 2.85$
   $\sigma^2 = \Sigma x^2 \cdot P(x) - \mu^2$
   $= 9.339 - (2.847)^2$
   $= 1.23$
   $\sigma = 1.11$

NOTE: Once $P(x=4) = .227$ was calculated, $P(x=5)$ could have been found by subtracting the other probabilities from 1.000. Calculating $P(x=5)$ separately and verifying the probabilities sum to 1.000, however, provides a check. In this instance, the sum of the probabilities is actually .999 due to rounding off. The values for μ and σ are reported to one more decimal point than usual to see how the answers check with the true values obtained below (without round-off error) using the special formulas that apply only to the binomial.

$\mu = n \cdot p$
$\quad = (5) \cdot (.57)$
$\quad = 2.85$
$\sigma^2 = n \cdot p \cdot (1-p)$
$\quad = (5) \cdot (.57) \cdot (.43)$
$\quad = 1.23$
$\sigma = 1.11$

And, to two decimal accuracy, the answers agree exactly. Of course, the answers always agree exactly when all decimal places are carried throughout to eliminate all round-off error.

7. binomial problem
let x = the number of California passengers wearing seat belts
$n = 500$
$p = .70$
a. $\mu = n \cdot p$
$\quad = (500) \cdot (.70)$
$\quad = 350.0$
$\sigma^2 = n \cdot p \cdot (1-p)$
$\quad = (500) \cdot (.70) \cdot (.30)$
$\quad = 105.0$
$\sigma = 10.2$

b. Unusual values are those outside $\mu \pm 2 \cdot \sigma$
$$350.0 \pm 2 \cdot (10.2)$$
$$350.0 \pm 20.4$$
$$329.6 \text{ to } 370.4$$

c. The result suggests that the campaign was successful and that the proportion is now greater than 70%, since 375 California drivers out of 500 is an unusual result if the true proportion is still 70%.

9. binomial problem
let x = the number of worker firings citing incompatibility
$n = 5$
$p = .17$
a. $x = 0$
$P(x) = [n!/x!(n-x)!] \cdot p^x \cdot (1-p)^{n-x}$
$\quad = [5!/0!5!] \cdot (.17)^0 \cdot (.83)^5$
$\quad = [1] \cdot (1.00) \cdot (.3939)$
$\quad = .394$

b. $x = 4$
$P(x) = [n!/x!(n-x)!] \cdot p^x \cdot (1-p)^{n-x}$
$\quad = [5!/4!1!] \cdot (.17)^4 \cdot (.83)^1$
$\quad = [5] \cdot (.0008352) \cdot (.83)$
$\quad = .00347$

c. $x = 5$
$P(x) = [n!/x!(n-x)!] \cdot p^x \cdot (1-p)^{n-x}$
$\quad = [5!/5!0!] \cdot (.17)^5 \cdot (.83)^0$
$\quad = [1] \cdot (.0001420) \cdot (1.00)$
$\quad = .000142$

d. x = at least 3
   for x=3, P(x) = $[n!/x!(n-x)!] \cdot p^x \cdot (1-p)^{n-x}$
                  = $[5!/3!2!] \cdot (.17)^3 \cdot (.83)^2$
                  = $[10] \cdot (.004913) \cdot (.6889)$
                  = .03385

   P(x≥3) = P(x=3) + P(x=4) + P(x=5)
          = .03385 + .00347 + .00014
          = .0375

Chapter 5

Normal Probability Distributions

5-2 The Standard Normal Distribution

1. The height of the rectangle is .2. Probability
corresponds to area, and the area of a rectangle
is (width)·(height).
$$P(3 < x < 5) = (\text{width})·(\text{height})$$
$$= (5-3)·(.2)$$
$$= (2)·(.2)$$
$$= .4$$

3. The height of the rectangle is .2. Probability
corresponds to area, and the area of a rectangle
is (width)·(height).
$$P(x < 4.5) = (\text{width})·(\text{height})$$
$$= (4.5-0)·(.2)$$
$$= (4.5)·(.2)$$
$$= .9$$

NOTE: The sketch is the key to exercises 5-28. It tells whether to add two Table A-2
probabilities, to subtract two Table A-2 probabilities, to subtract a Table A-2 probability from
.5000, to add a Table A-2 probability to .5000, etc., etc. It also often provides a check against
gross errors by indicating at a glance whether the final probability is less than or greater than
.5000. Remember that the symmetry of the normal curve implies two important facts:
   * There is always .5000 above and below the middle (i.e., at z = 0).
   * $P(-a < z < 0) = P(0 < z < a)$ for all values of "a."

5. $P(0 < z < 2.00)$
   $= .4772$

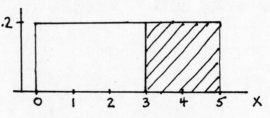

7. $P(z > 1.05)$
   $= P(z > 0) - P(0 < z < 1.05)$
   $= .5000 - .3531$
   $= .1469$

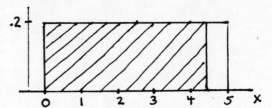

9. P(z > .50)
    = P(z > 0) - P(0 < z < .50)
    = .5000 - .1915
    = .3085

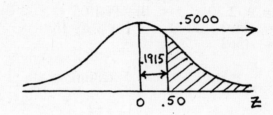

11. P(z > -1.09)
    = P(-1.09 < z < 0) + P(z > 0)
    = .3621 + .5000
    = .8621

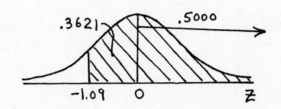

13. P(-1.00 < z < 1.00)
    = P(-1.00 < z < 0) + P(0 < z < 1.00)
    = .3413 + .3413
    = .6826

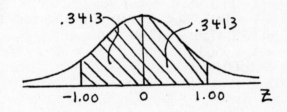

15. P(-1.15 < z < 2.60)
    = (-1.15 < z < 0) + P(0 < z < 2.60)
    = .3749 + .4953
    = .8702

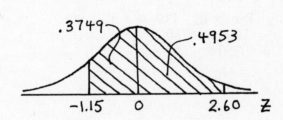

17. P(1.20 < z < 1.80)
    = P(0 < z < 1.80) - P(0 < z < 1.20)
    = .4641 - .3849
    = .0792

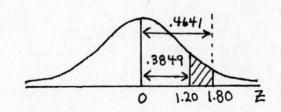

19. P(-2.30 < z < -1.05)
$\quad$ = P(-2.30 < z < 0) - P(-1.05 < z < 0)
$\quad$ = .4893 - .3531
$\quad$ = .1362

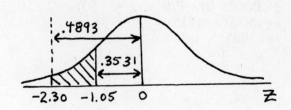

21. P(z > 0)
$\quad$ = .5000

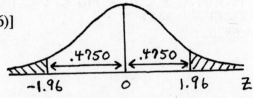

23. P(z < -1.96 or z > 1.96)
$\quad$ = 1 - P(-1.96 < z < 1.96)
$\quad$ = 1 - [P(-1.96 < z < 0) + P(0 < z < 1.96)]
$\quad$ = 1 - [.4750 + .4750]
$\quad$ = 1 - .9500
$\quad$ = .0500

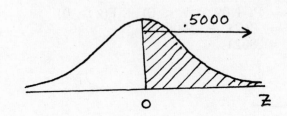

25. P(z > 1.96)
$\quad$ = P(z > 0) - P(0 < z < 1.96)
$\quad$ = .5000 - .4750
$\quad$ = .0250

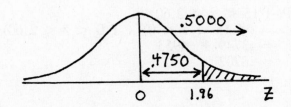

27. P(z < 2.33)
$\quad$ = P(z < 0) + P(0 < z < 2.33)
$\quad$ = .5000 + .4901
$\quad$ = .9901

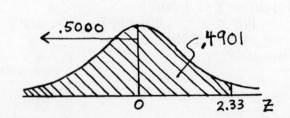

NOTE: The sketch is the key to exercises 29-36. It tells what probability is between 0 and the z score of interest (i.e. the area A to look up when reading Table A-2 "backwards." It also provides a check against gross errors by indicating at a glance whether a z score is above or below 0. Remember that the symmetry of the normal curve implies two important facts:
    * There is always .5000 above and below the middle (i.e., at z = 0).
    * $P(-a < z < 0) = P(0 < z < a)$ for all values of "a."

29. For P_{90}, A = .4000.
    The closest entry is A = .3997,
    for which z = 1.28.

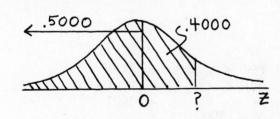

31. For the top 15%, A = .3500.
    The closest entry is A = .3508,
    for which z = 1.04 [positive
    since it is above the middle,
    where z = 0].

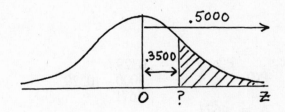

33. For the bottom 1%, A = .4900.
    The closest entry is A = .4901,
    for which z = -2.33 [negative
    since it is below the middle,
    where z = 0].

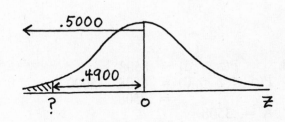

35. For the top 5%, A = .4500.
    The entry A = .4500 is at the top,
    for which z = 1.645.
    For the bottom 5%, A = .4500.
    The entry A = .4500 is at the top,
    for which z = -1.645 [negative
    since it is below the middle,
    where z = 0].

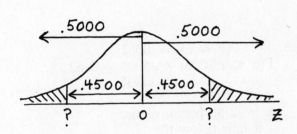

37. a. P(0 < z < a) = .4778
      A = .4778
      a = 2.01

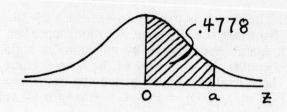

b. P(-b < z < b) = .7814
     P(0 < z < b) = .7814/2
                  = .3907
    A = .3907
    b = 1.23

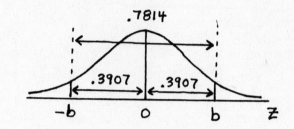

c.   P(z > c) = .0329
     P(0 < z < c) = .5000 - .0329
                  = .4671
    A = .4671
    c = 1.84

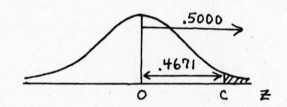

d.   P(z > d) = .8508
     P(d < z < 0) = .8508 - .5000
                  = .3508
    A = .3508
    d = -1.04 [negative since
       it falls below the
       middle, where z=0]

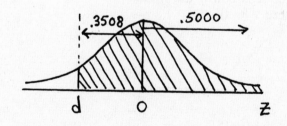

e.   P(z < e) = .0062
     P(e < z < 0) = .5000 - .0062
                  = .4938
    A = .4938
    e = -2.50 [negative since
       it falls below the
       middle, where z=0]

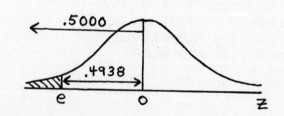

39.

x	$[2.7^{-x \cdot x/2}]/2.5$	y
-4	$(2.7)^{-8}/2.5$	.00014
-3	$(2.7)^{-4.5}/2.5$	.00458
-2	$(2.7)^{-2}/2.5$	.05487
-1	$(2.7)^{-.5}/2.5$	.24343
0	$(2.7)^{-0}/2.5$	.40000
1	$(2.7)^{-.5}/2.5$	.24343
2	$(2.7)^{-2}/2.5$	.05487
3	$(2.7)^{-4.5}/2.5$	.00458
4	$(2.7)^{-8}/2.5$	.00014

Approximate the area between z=0 and z=1 using a rectangle with width 1.0 and height (.40000 + .24343)/2 = .3217, the average of the heights of the curve at z=0 and z=1. The approximate area of (1.0)·(.3217) = .3217 compares well with the true area from Table A-2 of .3413.

5-3 Nonstandard Normal Distributions

NOTE: In each nonstandard normal distribution, x scores are converted to z scores using the formula $z = (x-\mu)/\sigma$ and rounded to two decimal places. The preceding formula may also be solved for x in terms of z to produce $x = \mu + z\sigma$. As in the previous section, drawing and labeling the sketch is the key to successful completion of the exercises.

1. $\mu = 100$
   $\sigma = 15$
   $P(100 < x < 127) = P(0 < z < 1.80)$
   $\qquad = .4641$

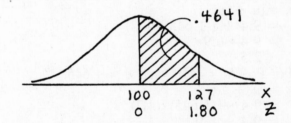

3. $\mu = 100$
   $\sigma = 15$
   $P(x > 76) = P(z > -1.60)$
   $\qquad\qquad = .4452 + .5000$
   $\qquad\qquad = .9452$

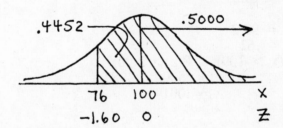

5. $\mu = 100$
   $\sigma = 15$
   For P_{90}, A = .4000.
   The closest entry is A = .3997,
   for which z = 1.28.
   $x = \mu + z\sigma$
   $\quad = 100 + (1.28) \cdot (15)$
   $\quad = 100 + 19.2$
   $\quad = 119.2$

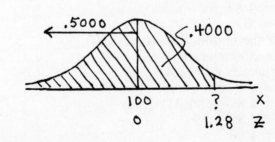

7. $\mu = 100$
$\sigma = 15$
For P_{20}, $A = .3000$.
The closest entry is $A = .2995$,
for which $z = -.84$.
$x = \mu + z\sigma$
$= 100 + (-.84) \cdot (15)$
$= 100 - 12.6$
$= 87.4$

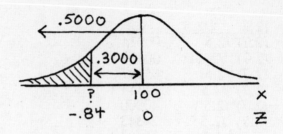

9. $\mu = 69.0$
$\sigma = 2.8$
$P(x < 64) = P(z < -1.79)$
$= .5000 - .4633$
$= .0367$

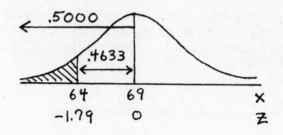

11. $\mu = 9.4$
$\sigma = 4.2$
For P_5, $A = .4500$ and $z = -1.645$.
$x = \mu + z\sigma$
$= 9.4 + (-1.645) \cdot (4.2)$
$= 9.4 - 6.9$
$= 2.5$
For P_{95}, $A = .4500$ and $z = 1.645$.
$x = \mu + z\sigma$
$= 9.4 + (1.645) \cdot (4.2)$
$= 9.4 + 6.9$
$= 16.3$

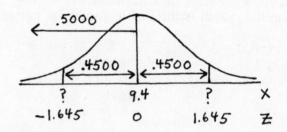

13. $\mu = 98.2$
$\sigma = .62$
$P(x > 100) = P(z > 2.90)$
$= .5000 - .4981$
$= .0019$

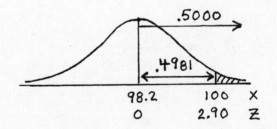

15. $\mu = 4.89$
$\sigma = .63$
For the bottom 8%, $A = .4200$.
The closest entry is $A = .4207$,
for which $z = -1.41$.
$x = \mu + z\sigma$
$= 4.89 + (-1.41) \cdot (.63)$
$= 4.89 - .89$
$= 4.00$

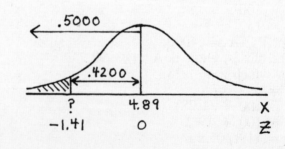

17. $\mu = 3.80$
$\sigma = .95$
$P(x > 4.00) = P(z > .21)$
$= .5000 - .0832$
$= .4168$

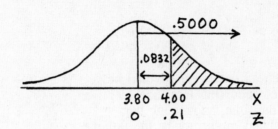

19. $\mu = 99.56$
$\sigma = 25.84$
$P(110 < x < 150) = P(.40 < z < 1.95)$
$= .4744 - .1554$
$= .3190$

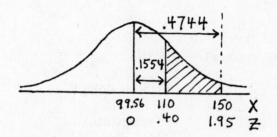

21. $\mu = 58.84$
$\sigma = 15.94$
$P(x < 27.00) = P(z < -2.00)$
$= .5000 - .4772$
$= .0228$

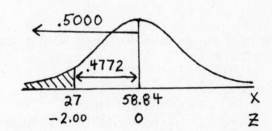

23. $\mu = 22.83$
$\sigma = 8.55$
$P(4.02 < x < 22.83) = P(-2.20 < z < 0)$
$= .4861$

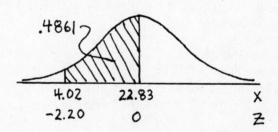

25. $\mu = 8.0$
$\sigma = 2.6$
$P(6.0 < x < 7.0) = P(-.77 < z < -.38)$
$= .2794 - .1480$
$= .1314$
$(.1314) \cdot (600) = 78.8$

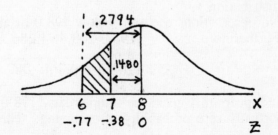

27. $\mu = 268$
 $\sigma = 15$
 $P(x > 308) = P(z > 2.67)$
 $\qquad\qquad = .5000 - .4962$
 $\qquad\qquad = .0038$
Such an occurrence would be
rare, occurring less than 1%
of the time, suggesting her
husband might not be the father.
On the other hand, in a small city
of 100,000 people there should be about
$(.0038) \cdot (100,000) = 380$ people who were born under just such circumstances.

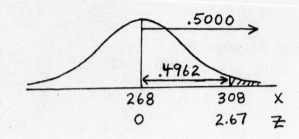

29. $\mu = 5.67$
 $\sigma = .0700$
 $P(\text{acceptance}) = P(5.50 < x < 5.80)$
 $\qquad\qquad = P(-.2.43 < z < 1.86)$
 $\qquad\qquad = .4925 + .4686$
 $\qquad\qquad = .9611$
 $P(\text{rejection}) = 1 - P(\text{acceptance})$
 $\qquad\qquad = 1.0000 - .9611$
 $\qquad\qquad = .0389$

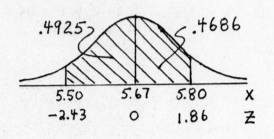

31. preliminary values: $n = 40$, $\Sigma x = 8018.4$, $\Sigma x^2 = 1,607,371.66$
 $\overline{x} = \Sigma x/n$
 $\qquad = 8018.4/40$
 $\qquad = 200.46$
 $s^2 = [n(\Sigma x^2) - (\Sigma x)^2]/[n(n-1)]$
 $\qquad = [40(1,607,371.66) - (8018.4)^2]/[40(39)]$
 $\qquad = (127.84)/1560$
 $\qquad = .0819$
 $s = .286$
 a. $3/40 = .075$ or 7.5%
 b. $\overline{x} = 200.46$
 c. $s = .29$
 d. $\mu = 200.46$
 $\sigma = .286$
 $P(x > 201.0) = P(z > 1.89)$
 $\qquad\qquad = .5000 - .4706$
 $\qquad\qquad = .0294$

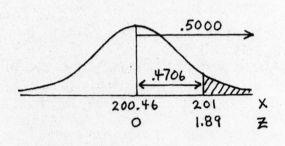

NOTE: The observed proportion in part (a) of .0750 differs from this value predicted assuming a normal distribution. This may suggest that the times do not follow a normal distribution.
 e. No observations occurred below 198.0 or above 202.0.
 Furthermore, $z_{198.0} = (198.0-200.46)/.286$
 $\qquad\qquad = -8.60$
 $z_{202.0} = (202.0-200.46)/.286$
 $\qquad\qquad = 5.38$

It appears that given the sample data, 198.0 and 202.0 would be extremely rare events with virtually zero probability of occurring. The specifications seem to be met.

33. The cut-off points are P_{90}, P_{70}, P_{30}, P_{10}.
$\mu = 50$
$\sigma = 10$
For P_{90}, $A = .4000$.
   The closest entry is $A = .3997$, for which $z = 1.28$
   $x = \mu + z\sigma$
     $= 50 + (1.28) \cdot (10)$
     $= 50 + 12.8$
     $= 62.8$

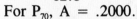

For P_{70}, $A = .2000$.
   The closest entry is $A = .1985$, for which $z = .52$
   $x = \mu + z\sigma$
     $= 50 + (.52) \cdot (10)$
     $= 50 + 5.2$
     $= 55.2$

For P_{30}, $A = .2000$.
   The closest entry is $A = .1985$, for which $z = -.52$
   $x = \mu + z\sigma$
     $= 50 + (-.52) \cdot (10)$
     $= 50 - 5.2$
     $= 44.8$

For P_{10}, $A = .4000$.
   The closest entry is $A = .3997$, for which $z = -1.28$
   $x = \mu + z\sigma$
     $= 50 + (-1.28) \cdot (10)$
     $= 50 - 12.8$
     $= 37.2$

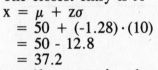

The grades are assigned as follows.
   A: at least 62.8
   B: at least 55.2 and less than 62.8
   C: at least 44.8 and less than 55.2
   D: at least 37.2 and less than 44.8
   F: less than 37.2

35. $\mu = 36.2$
$\sigma = 3.8$
For the top $200/4830 = .0414$, $A = .4586$.
The closest entry is $A = .4582$, for which $z = -1.73$.
NOTE: The top runners have the <u>lower</u> times (i.e., they occur below the mean and have negative z scores.

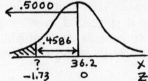

$x = \mu + z\sigma$
   $= 36.2 + (-1.73) \cdot (3.8)$
   $= 36.2 - 6.6$
   $= 29.6$

37. First, find the z score for which $P(-a < z < a) = 2/3$ -- i.e., for which
$$P(0 < z < a) = 1/3$$
$$= .3333$$
   The closest entry is $A = .3340$, for which the score is $.97$.
Now use the above z score and the condition that the difference between x and μ is 30 to solve
   $z = (x-\mu)/\sigma$ for σ to get $\sigma = (x-\mu)/z$
$$= 30/.97$$
$$= 30.9$$

5-4 The Central Limit Theorem

NOTE: When working with individual scores (i.e., making a statement about a single x scores from the original distribution), convert x to z using the mean and standard deviation of the x's and $z = (x-\mu)/\sigma$.

When working with a sample of n scores (i.e., making a statement about $\bar{x}$), convert $\bar{x}$ to z using the mean and standard deviation of the $\bar{x}$'s and $z = (\bar{x}-\mu_{\bar{x}})/\sigma_{\bar{x}}$.

1. a. normal distribution
  $\mu = 100$
  $\sigma = 15$
  $P(100 < x < 103) = P(0 < z < .20)$
  $\qquad\qquad\qquad = .0793$

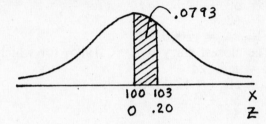

b. normal distribution, since the original distribution is so
  $\mu_{\bar{x}} = \mu = 100$
  $\sigma_{\bar{x}} = \sigma/\sqrt{n} = 15/\sqrt{25} = 3$
  $P(100 < \bar{x} < 103) = P(0 < z < 1.00)$
  $\qquad\qquad\qquad = .3413$

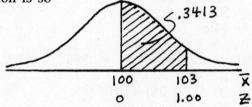

3. normal distribution, since the original distribution is so
  $\mu_{\bar{x}} = \mu = 9.43$
  $\sigma_{\bar{x}} = \sigma/\sqrt{n} = 4.17/\sqrt{12} = 1.204$
  $P(10.0 < \bar{x} < 12.0) = P(.47 < z < 2.13)$
  $\qquad\qquad\qquad = .4834 - .1808$
  $\qquad\qquad\qquad = .3026$

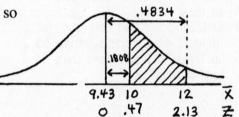

IMPORTANT NOTE: After calculating $\sigma_{\bar{x}}$, <u>STORE IT</u> in the calculator to recall it with total accuracy whenever it is needed in subsequent calculations. <u>DO NOT</u> write it down on paper rounded off (even to several decimal places) and then re-enter it in the calculator whenever it is needed. This avoids both round off errors and recopying errors.

5. normal distribution, since the original distribution is so
  $\mu_{\bar{x}} = \mu = 34.8$
  $\sigma_{\bar{x}} = \sigma/\sqrt{n} = 7.02/\sqrt{36} = 1.17$
  $P(34.8 < \bar{x} < 37.0) = P(0 < z < 1.88)$
  $\qquad\qquad\qquad = .4699$

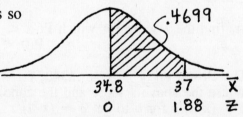

7. normal distribution, by the Central Limit Theorem
$\mu_{\bar{x}} = \mu = 10.7$
$\sigma_{\bar{x}} = \sigma/\sqrt{n} = 11.2/\sqrt{42} = 1.728$
$P(\bar{x} < 12.0) = P(z < .75)$
$= .5000 + .2734$
$= .7734$

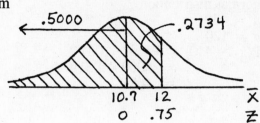

9. normal distribution, by the Central Limit Theorem
$\mu_{\bar{x}} = \mu = 13.0$
$\sigma_{\bar{x}} = \sigma/\sqrt{n} = 7.9/\sqrt{35} = 1.335$
$P(\bar{x} > 15) = P(z > 1.50)$
$= .5000 - .4332$
$= .0668$

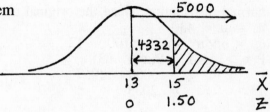

11. normal distribution, by the Central Limit Theorem
[since $31/630 < .05$, the finite population correction factor is not needed]
$\mu_{\bar{x}} = \mu = 2.97$
$\sigma_{\bar{x}} = \sigma/\sqrt{n} = .60/\sqrt{31} = .108$
$P(3.00 < \bar{x} < 3.10) = P(.28 < z < 1.21)$
$= .3869 - .1103$
$= .2766$

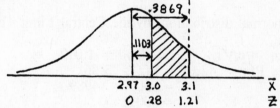

13. normal distribution, since the original distribution is so
$\mu_{\bar{x}} = \mu = 8.5$
$\sigma_{\bar{x}} = \sigma/\sqrt{n} = 3.962/\sqrt{36} = .66$
$P(7.0 < \bar{x} < 10.0) = P(-2.27 < z < 2.27)$
$= .4884 + .4884$
$= .9768$

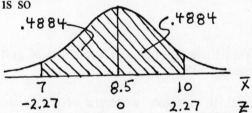

15. normal distribution, by the Central Limit Theorem
$\mu_{\bar{x}} = \mu = 3.54$
$\sigma_{\bar{x}} = \sigma/\sqrt{n} = .96/\sqrt{80} = .107$
$P(3.50 < \bar{x} < 3.60) = P(-.37 < z < .56)$
$= .1443 + .2123$
$= .3566$

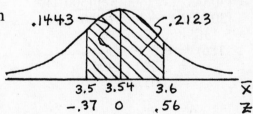

17. a. normal distribution

$\mu = 430$

$\sigma = 120$

$P(x > 440) = P(z > .08)$
$= .5000 - .0319$
$= .4681$

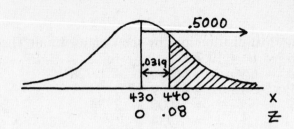

b. normal distribution, since the original distribution is so

$\mu_{\bar{x}} = \mu = 430$

$\sigma_{\bar{x}} = \sigma/\sqrt{n} = 120/\sqrt{100} = 12$

$P(\bar{x} > 440) = P(z > .83)$
$= .5000 - .2967$
$= .2033$

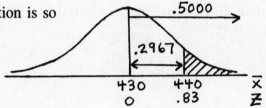

c. Not necessarily; 20% of the time 100 students with no special training can be expected to achieve a mean score of 440 or higher.

19. normal distribution, by the Central Limit Theorem

$\mu_{\bar{x}} = \mu = 38.9$

$\sigma_{\bar{x}} = \sigma/\sqrt{n} = 12.4/\sqrt{150} = 1.012$

a. $P(\bar{x} > 42.0) = P(z > 3.06)$
$= .5000 - .4989$
$= .0011$

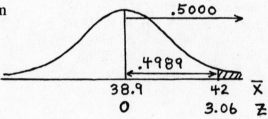

b. Yes; there is only a .1% chance of getting such a high sample mean if the true population mean is 38.9.

21. normal distribution, since the original distribution is so

[since 32/500 > .05, the finite population correction factor must be used]

$\mu_{\bar{x}} = \mu = 173$

$\sigma_{\bar{x}} = [\sigma/\sqrt{n}] \cdot \sqrt{(N-n)/(N-1)}$
$= [30/\sqrt{32}] \cdot \sqrt{(500-32)/(500-1)}$
$= [30/\sqrt{32}] \cdot \sqrt{(468)/(499)}$
$= 5.136$

$P(\bar{x} > 186) = P(z > 2.53)$
$= .5000 - .4943$
$= .0057$

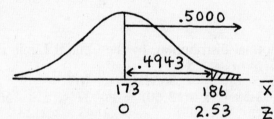

23. normal distribution, since the original distribution appears so
    [since 15/350 < .05, the finite population correction factor is not needed]
    NOTE: From the boxplot we determine,
    $\bar{x} = 50$
    $s = 22.4$ In a normal distribution, $P(-.67 < z < .67) = .5000$ from Table A-2.
    The hinges indicate that $P(35 < x < 65) = .5000$.
    Solving $z = (x-\bar{x})/s$ for s yields $s = (x-\bar{x})/z$
    $= (65-50)/.67$
    $= 15/.67$
    $= 22.4$

Using $\bar{x}$ and s to estimate μ and σ
$\mu_{\bar{x}} = \mu = 50$
$\sigma_{\bar{x}} = \sigma/\sqrt{n} = 22.4/\sqrt{15} = 5.78$
$P(\bar{x} > 55) = P(z > .86)$
$= .5000 - .3051$
$= .1949$

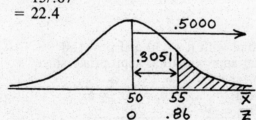

5-5 Normal Distribution as Approximation to Binomial Distribution

1. a. Table A-1 with $n = 12$ and $p = .50$
      $P(x = 8) = .121$

   b. normal approximation appropriate since
      $np = 12(.50) = 6 \geq 5$
      $n(1-p) = 12(.50) = 6 \geq 5$
      $\mu = np = 12(.50) = 6$
      $\sigma = \sqrt{np(1-p)} = \sqrt{12(.50)(.50)} = 1.732$
      $P(x = 8) = P_c(7.5 < x < 8.5)$
      $= P(.87 < z < 1.44)$
      $= .4251 - .3078$
      $= .1173$

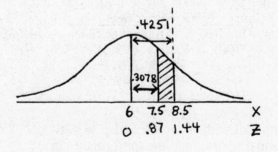

IMPORTANT NOTE: As in the previous section, store σ in the calculator so that it may be recalled with complete accuracy whenever it is needed in subsequent calculations. As before, P(E) represents the probability of an event E; this manual uses $P_c(E)$ to represent the probability of an event E with the continuity correction applied.

3. a. Table A-1 with $n = 10$ and $p = .80$
      $P(x \geq 8) = P(8) + P(9) + P(10)$
      $= .302 + .268 + .107$
      $= .677$

   b. normal approximation not appropriate since
      $np = 10(.80) = 8 \geq 5$
      $n(1-p) = 10(.20) = 2 < 5$

5. binomial with n = 100 and p = .50
   normal approximation appropriate since
   $$np = 100(.50) = 50 \geq 5$$
   $$n(1-p) = 100(.50) = 50 \geq 5$$
   $$\mu = np = 100(.50) = 50$$
   $$\sigma = \sqrt{np(1-p)} = \sqrt{100(.50)(.50)} = 5$$
   $$P(x \geq 60) = P_c(x > 59.5)$$
   $$= P(z > 1.90)$$
   $$= .5000 - .4713$$
   $$= .0287$$

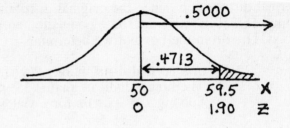

7. binomial with n = 50 and p = .50
   normal approximation appropriate since
   $$np = 50(.50) = 25 \geq 5$$
   $$n(1-p) = 50(.50) = 25 \geq 5$$
   $$\mu = np = 50(.50) = 25$$
   $$\sigma = \sqrt{np(1-p)} = \sqrt{50(.50)(.50)} = 3.536$$
   $$P(x \geq 30) = P_c(x > 29.5)$$
   $$= P(z > 1.27)$$
   $$= .5000 - .3980$$
   $$= .1020$$

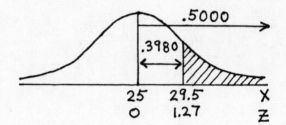

9. binomial with n = 1000 and p = .66
   normal approximation appropriate since
   $$np = 1000(.66) = 660 \geq 5$$
   $$n(1-p) = 1000(.34) = 340 \geq 5$$
   $$\mu = np = 1000(.66) = 660$$
   $$\sigma = \sqrt{np(1-p)} = \sqrt{1000(.66)(.34)} = 14.980$$
   $$P(x \geq 700) = P_c(x > 699.5)$$
   $$= P(z > 2.64)$$
   $$= .5000 - .4959$$
   $$= .0041$$

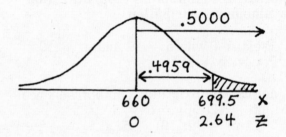

11. binomial with n = 500 and p = .26
    normal approximation appropriate since
    $$np = 500(.26) = 130 \geq 5$$
    $$n(1-p) = 500(.74) = 370 \geq 5$$
    $$\mu = np = 500(.26) = 130$$
    $$\sigma = \sqrt{np(1-p)} = \sqrt{500(.26)(.74)} = 9.808$$
    $$P(125 \leq x \leq 150) = P_c(124.5 < x < 150.5)$$
    $$= P(-.56 < z < 2.09)$$
    $$= .2132 + .4817$$
    $$= .6940$$

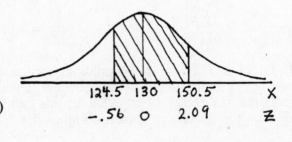

13. binomial with n = 2500 and p = .24
    normal approximation appropriate since
    $$np = 2500(.24) = 600 \geq 5$$
    $$n(1-p) = 2500(.76) = 1900 \geq 5$$
    $$\mu = np = 2500(.24) = 600$$
    $$\sigma = \sqrt{np(1-p)} = \sqrt{2500(.24)(.76)} = 21.354$$
    $$P(x > 650) = P_c(x > 650.5)$$
    $$= P(z > 2.36)$$
    $$= .5000 - .4909$$
    $$= .0091$$

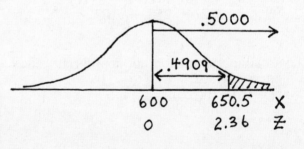

15. binomial with n = 600 and p = .35
   normal approximation appropriate since
   $np = 600(.35) = 210 \geq 5$
   $n(1-p) = 600(.65) = 390 \geq 5$
   $\mu = np = 600(.35) = 210$
   $\sigma = \sqrt{np(1-p)} = \sqrt{600(.35)(.65)} = 11.683$
   $P(x \geq 210) = P_c(x > 209.5)$
   $\qquad = P(z > -.04)$
   $\qquad = .0160 + .5000$
   $\qquad = .5160$

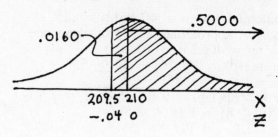

17. binomial with n = 320 and p = .75
   normal approximation appropriate since
   $np = 320(.75) = 240 \geq 5$
   $n(1-p) = 320(.25) = 80 \geq 5$
   $\mu = np = 320(.75) = 240$
   $\sigma = \sqrt{np(1-p)} = \sqrt{320(.75)(.25)} = 7.746$
   $P(x > 250) = P_c(x > 250.5)$
   $\qquad = P(z > 1.36)$
   $\qquad = .5000 - .4131$
   $\qquad = .0869$

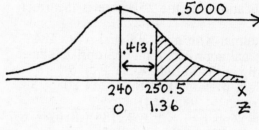

19. binomial with n = 100 and p = .20
   normal approximation appropriate since
   $np = 100(.20) = 20 \geq 5$
   $n(1-p) = 100(.80) = 80 \geq 5$
   $\mu = np = 100(.20) = 20$
   $\sigma = \sqrt{np(1-p)} = \sqrt{100(.20)(.80)} = 4$
   $P(x \leq 17) = P_c(x < 17.5)$
   $\qquad = P(z < -.625)$
   $\qquad = P(z < -.62)$ or $P(z < -.63)$
   $\qquad = .5000 - .2324$ or $.5000 - .2357$
   $\qquad = .2676$ or $.2643$

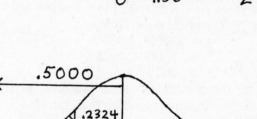

21. binomial with n = 400 and p = .16
   normal approximation appropriate since
   $np = 400(.16) = 64 \geq 5$
   $n(1-p) = 400(.84) = 336 \geq 5$
   $\mu = np = 400(.16) = 64$
   $\sigma = \sqrt{np(1-p)} = \sqrt{400(.16)(.84)} = 7.332$
   $P(x \geq 100) = P_c(x > 99.5)$
   $\qquad = P(z > 4.84)$
   $\qquad = .5000 - .4999$ [see note at top of Table A-2]
   $\qquad = .0001$

No; it does not seem plausible that the state arrests women at the 16% rate.

23. binomial with n = 40 and p = .25
    normal approximation appropriate since
    $\quad np = 40(.25) = 10 \geq 5$
    $\quad n(1-p) = 40(.75) = 30 \geq 5$
    $\mu = np = 40(.25) = 10$
    $\sigma = \sqrt{np(1-p)} = \sqrt{40(.25)(.75)} = 2.739$
    $P(x \leq 8) = P_c(x < 8.5)$
    $\qquad\qquad = P(z < -.55)$
    $\qquad\qquad = .5000 - .2088$
    $\qquad\qquad = .2912$

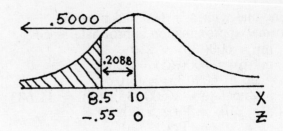

The result is well within the reasonable expectation for a 25% rate does not give any indication that the 25% rate is incorrect.

25. binomial with n = 300 and p = .184
    normal approximation appropriate since
    $\quad np = 300(.184) = 55.2 \geq 5$
    $\quad n(1-p) = 300(.816) = 244.80 \geq 5$
    $\mu = np = 300(.184) = 55.2$
    $\sigma = \sqrt{np(1-p)} = \sqrt{300(.184)(.816)} = 6.711$
    $P(x \geq 72) = P_c(x > 71.5)$
    $\qquad\qquad = P(z > 2.43)$
    $\qquad\qquad = .5000 - .4925$
    $\qquad\qquad = .0075$

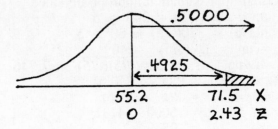

Under the conditions of the problem, getting 72 or more smokers by chance alone is very unlikely. It appears either that the 18.4% figure is incorrect or that there is something wrong with the sample.

27. binomial with n = 400 and p = .45
    normal approximation appropriate since
    $\quad np = 400(.45) = 180 \geq 5$
    $\quad n(1-p) = 400(.55) = 220 \geq 5$
    $\mu = np = 400(.45) = 180$
    $\sigma = \sqrt{np(1-p)} = \sqrt{400(.45)(.55)} = 9.950$
    $P(x \geq 177) = P_c(x > 176.5)$
    $\qquad\qquad = P(z > -.35)$
    $\qquad\qquad = .1368 + .5000$
    $\qquad\qquad = .6368$

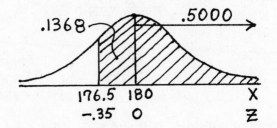

29. binomial with n = 15 and p = .4
    a. using Table A-1

x	P(x)
5	.186
6	.207
7	.177
8	.118
9	.061
10	.024
11	.007
12	.002
13	0+
14	0+
15	0+
	.782

b. using the binomial formula $P(x) = [n!/x!(n-x)!]p^x(1-p)^{n-x}$

x	P(x)	
5	$3003(.4)^5(.6)^{10}$	= .1859378
6	$5005(.4)^6(.6)^9$	= .2065976
7	$6435(.4)^7(.6)^8$	= .1770837
8	$6435(.4)^8(.6)^7$	= .1880558
9	$5005(.4)^9(.6)^6$	= .0612142
10	$3003(.4)^{10}(.6)^5$	= .0244856
11	$1365(.4)^{11}(.6)^4$	= .0074190
12	$455(.4)^{12}(.6)^3$	= .0016489
13	$105(.4)^{13}(.6)^2$	= .0002537
14	$15(.4)^{14}(.6)^1$	= .0000242
15	$1(.4)^{15}(.6)^0$	= .0000011
		.7827215

c. binomial with n = 15 and p = .4
   normal approximation appropriate since
   $np = 15(.4) = 6 \geq 5$
   $n(1-p) = 15(.6) = 9 \geq 5$
   $\mu = np = 15(.4) = 6$
   $\sigma = \sqrt{np(1-p)} = \sqrt{15(.4)(.6)} = 1.897$
   $P(x \geq 5) = P_c(x > 4.5)$
   $= P(z > -.79)$
   $= .2852 + .5000$
   $= .7852$

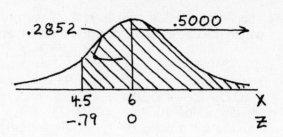

Of the three answers (a) $P(x \geq 5) = .782$
  (b) $P(x \geq 5) = .7827215$
  (c) $P(x \geq 5) = .7852$,
answer (b) is the one closest to the correct answer.

31. Letting x represent the number of persons with advanced reservations who <u>do</u> show up for a flight, the problem is a binomial one with n unknown and p = .93. The airline wants to find the largest n for which $P(x \leq 250) = .95$. The problem may be solved in two different manners -- (a) increasing n and calculating $P(x \leq 250)$ until $P(x \leq 250)$ drops below .95, (b) solving directly for the value of n at which $P(x \leq 250)$ equal .95.

(a) The following table summarizes the procedure for finding $P(x \leq 250) = P_c(x < 250.5)$, where p = .93.

n	$\mu = np$	$\sigma = \sqrt{np(1-p)}$	$z = (250.5-\mu)/\sigma$	$P(x<250.5)$
260	241.8	4.11	2.11	.9826
261	242.7	4.12	1.89	.9706
262	243.7	4.13	1.66	.9515
263	244.6	4.14	1.43	.9236
264	245.5	4.15	1.20	.8849
265	246.5	4.15	.98	.8365
266	247.4	4.16	.75	.7734
267	248.3	4.17	.53	.7019
268	249.2	4.18	.30	.6179
269	250.2	4.18	.08	.5319
270	251.1	4.19	-.14	.4443
271	252.0	4.20	-.36	.3594
272	253.0	4.21	-.58	.2810
273	253.9	4.22	-.80	.2119
274	254.8	4.22	-1.02	.1539
275	255.8	4.23	-1.24	.1075
276	256.7	4.24	-1.46	.0721
277	257.6	4.25	-1.67	.0475

And so for n = 262 there is a 95% probability that everyone who shows up with advanced reservations will have a seat. For n > 262 the probability falls below that acceptable level.

(b) The z value below which 95% of the probability occurs is 1.645. Solving to find the n for which 250.5 or fewer persons with advanced reservations will show up 95% of the time,

$$(x-\mu)/\sigma = z$$
$$(250.5 - .93n)/\sqrt{(.93)(.07)n} = 1.645$$
$$(250.5) - .93n = 1.645 \cdot \sqrt{(.93)(.07)n}$$
$$62750.25 - 465.93n + .8649n^2 = .1762n$$
$$.8649n^2 - 466.1062n + 62750.25 = 0$$

Solving for n using the quadratic formula,

$$n = [466.1062 \pm \sqrt{(-466.1062)^2 - 4(.8649)(62750.25)}]/2(.8649)$$
$$= [466.1062 \pm 12.8136]/1.7298$$
$$= 453.2925/1.7298 \quad or \quad 478.9198/1.7298$$
$$= 262.05 \quad or \quad 276.86$$

And so z = 1.645 when n = 262.05. If n > 262.05 then z < 1.645 and the probability is less than 95%, so we round down to n = 262. Note that z = -1.645 when n = 276.86 [and that extraneous root was introduced when both sides of the equation were squared].

33. binomial with n = 200 and p = P(a woman is taller than 66 inches)
    = [not given directly, must be calculated]
adult female heights follow a normal distribution with $\mu = 63.6$ and $\sigma = 2.5$

$$P(x > 66) = P(z > .96)$$
$$= .5000 - .3315$$
$$= .1685$$

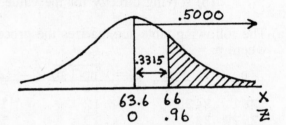

binomial with n = 200 and p = .1685
normal approximation appropriate since
$$np = 200(.1685) = 33.7 \geq 5$$
$$n(1-p) = 200(.8315) = 166.3 \geq 5$$
$$\mu = np = 200(.1685) = 33.7$$
$$\sigma = \sqrt{np(1-p)} = \sqrt{200(.1685)(.8315)} = 5.294$$
$$P(x \geq 40) = P_c(x > 39.5)$$
$$= P(z > 1.10)$$
$$= .5000 - .3643$$
$$= .1357$$

Review Exercises

1. normal distribution with $\mu = 106$ and $\sigma = 3$

   a. P(x < 100) = P(z < -2.00)

   $\phantom{a. P(x < 100)} = .5000 - .4772$

   $\phantom{a. P(x < 100)} = .0228$

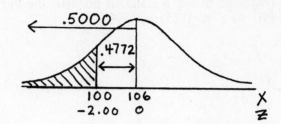

   b. P(x > 110) = P(z > 1.33)

   $ = .5000 - .4082$

   $ = .0918$

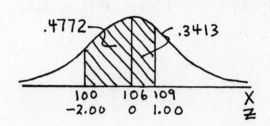

   c. P(x < 112) = P(z < 2.00)

   $\phantom{c. P(x < 112)} = .5000 + .4772$

   $\phantom{c. P(x < 112)} = .9772$

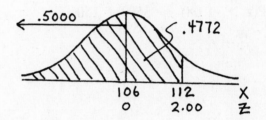

   d. P(100 < x < 109) = P(-2.00 < z < 1.00)

   $\phantom{d. P(100 < x < 109)} = .4772 + .3413$

   $\phantom{d. P(100 < x < 109)} = .8185$

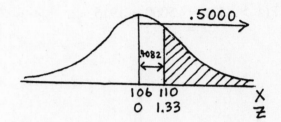

   e. P(110 < x < 115) = P(1.33 < z < 3.00)

   $\phantom{e. P(110 < x < 115)} = .4987 - .4082$

   $\phantom{e. P(110 < x < 115)} = .0905$

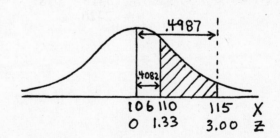

3. normal distribution with $\mu = 0$ and $\sigma = 1$
   [Because this is a standard normal, the variable is already expressed in terms of z scores.]
   a. $P(0 < z < 1.25) = .3994$

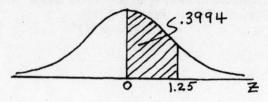

   b. $P(z > .50) = .5000 - .1915$
   $= .3085$

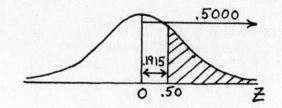

   c. $P(z > -1.08) = .3599 + .5000$
   $= .8599$

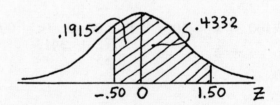

   d. $P(-.50 < z < 1.50) = .1915 + .4332$
   $= .6247$

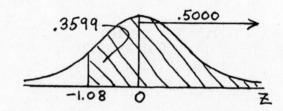

   e. $P(-1.00 < z < -.25) = .3413 - .0987$
   $= .2426$

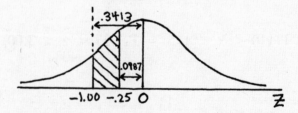

5. normal distribution, by the Central Limit Theorem

$\mu_{\bar{x}} = \mu = 41182$

$\sigma_{\bar{x}} = \sigma/\sqrt{n} = 19990/\sqrt{125} = 1787.96$

$P(\bar{x} > 40000) = P(z > -.66)$
$\qquad\qquad\quad = .2454 + .5000$
$\qquad\qquad\quad = .7454$

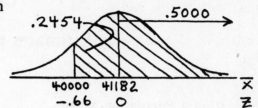

No additional assumptions about the data are necessary -- the Central Limit Theorem applies to <u>any</u> distribution with a given μ and σ.

7. binomial with n = 225 and p = .08
normal approximation appropriate since

$np = 225(.08) = 18 \geq 5$

$n(1-p) = 225(.92) = 207 \geq 5$

$\mu = np = 225(.08) = 18$

$\sigma = \sqrt{np(1-p)} = \sqrt{225(.08)(.92)} = 4.07$

$P(x \geq 20) = P_c(x > 19.5)$
$\qquad\qquad = P(z > .37)$
$\qquad\qquad = .5000 - .1443$
$\qquad\qquad = .3557$

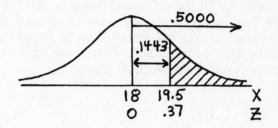

9. binomial with n = 200 and p = .10
normal approximation appropriate since

$np = 200(.10) = 20 \geq 5$

$n(1-p) = 200(.90) = 180 \geq 5$

$\mu = np = 200(.10) = 20$

$\sigma = \sqrt{np(1-p)} = \sqrt{200(.10)(.90)} = 4.24$

a. $P(x = 18) = P_c(17.5 < x < 18.5)$
$\qquad\qquad\quad = P(-.59 < z < -.35)$
$\qquad\qquad\quad = .2224 - .1368$
$\qquad\qquad\quad = .0856$

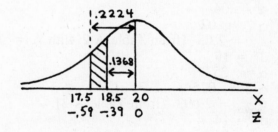

b. $P(x < 25) = P_c(x < 24.5)$
$\qquad\qquad = P(z < 1.06)$
$\qquad\qquad = .5000 + .3554$
$\qquad\qquad = .8554$

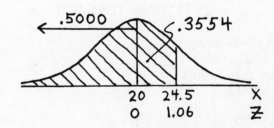

Chapter 6

Estimates and Sample Sizes

6-2 Estimating a Population Mean

IMPORTANT NOTE: This manual uses the following conventions.

(1) The designation "df" stands for "degrees of freedom."

(2) Since the t value depends on both the degrees of freedom and the probability lying beyond it, double subscripts are used to identify points on t distributions. The t distribution with 15 degrees of freedom and .025 beyond it, for example, is designated $t_{15,.025} = 2.132$.

(3) When df $\geq$ 30 the difference between the t and z distributions is negligible and Table A-3 uses one final row of z values to cover all such cases. Consequently, the z scores for certain "popular" α and $\alpha/2$ values may be found by reading Table A-3 "frontwards" instead of reading Table A-2 "backwards." This is not only easier but also more accurate, since Table A-3 includes one more decimal place.

1. a. $\alpha = .05$
   $\alpha/2 = .025$
   $z_{.025} = 1.960$ [from Table A-3]
   b. $\alpha = .02$
   $\alpha/2 = .01$
   $z_{.01} = 2.327$ [from Table A-3]
   c. $\alpha = .04$
   $\alpha/2 = .02$
   $z_{.02} = 2.05$ [from Table A-2 with A $= .4800$ (or closest entry)]
   d. df $= 19$
   $\alpha = .05$
   $\alpha/2 = .025$
   $t_{19,.025} = 2.093$ [from Table A-3]
   e. df $= 14$
   $\alpha = .01$
   $\alpha/2 = .005$
   $t_{14,.005} = 2.971$ [from Table A-3]

3. σ known, use $\underline{z}$
   $E = z_{.005} \cdot \sigma/\sqrt{n}$
   $= 2.575 \cdot 40/\sqrt{25}$
   $= 20.6$

5. $n > 30$, use $\underline{z}$ (with s for σ)
   $E = z_{.005} \cdot \sigma/\sqrt{n}$
   $= 2.575 \cdot 15/\sqrt{100}$
   $= 3.8625$

7. σ known, use $\underline{z}$
   $\bar{x} \pm z_{.025} \cdot \sigma/\sqrt{n}$
   $70.4 \pm 1.960 \cdot 5/\sqrt{36}$
   70.4 ± 1.6
   $68.8 < \mu < 72.0$

9. $\underline{n} > 30$, use $\underline{z}$ (with s for σ)
$\bar{x} \pm z_{.005} \cdot \sigma / \sqrt{n}$
$5.15 \pm 2.575 \cdot 1.68 / \sqrt{4400}$
$5.15 \pm .07$
$5.08 < \mu < 5.22$

11. $\underline{n} \leq 30$ and σ unknown, use t
$\bar{x} \pm t_{19,.025} \cdot s / \sqrt{n}$
$2.40 \pm 2.093 \cdot 1.30 / \sqrt{20}$
$2.40 \pm .61$
$1.79 < \mu < 3.01$

13. $E = 1.5$
$\alpha = .02$
$n = [z_{.01} \cdot \sigma / E]^2$
$= [2.327 \cdot 15 / 1.5]^2$
$= 541.49$ rounded up to 542

15. $\underline{n} > 30$, use $\underline{z}$ (with s for σ)
$\bar{x} \pm z_{.005} \cdot \sigma / \sqrt{n}$
$5.622 \pm 2.575 \cdot 0.68 / \sqrt{50}$
$5.622 \pm .025$
$5.597 < \mu < 5.647$
While this confidence interval does not include the value 5.670, it does not contradict the mint's claim that quarters are produced at that weight. The mint is making a claim about the weight of new (i.e., uncirculated) quarters, but the quarters in the sample were taken from general circulation and had experienced wear. This problem reinforces the importance of making certain that the sample selected for observation is representative of the population of interest.

17. $E = 2.0$
$\alpha = .05$
$n = [z_{.025} \cdot \sigma / E]^2$
$= [1.960 \cdot 21.2 / 2.0]^2$
$= 431.64$ rounded up to 432

19. $\underline{n} > 30$, use $\underline{z}$ (with s for σ)
$\bar{x} \pm z_{.005} \cdot \sigma / \sqrt{n}$
$27.44 \pm 2.575 \cdot 12.46 / \sqrt{62}$
27.44 ± 4.07
$23.37 < \mu < 31.51$
No; since we are 99% certain that the true mean is less than 35, we are 99% certain that the facility will not be overburdened.

21. a. $\underline{n} > 30$, use $\underline{z}$ (with s for σ)
$\bar{x} \pm z_{.005} \cdot \sigma / \sqrt{n}$
$41.8 \pm 2.575 \cdot 16.7 / \sqrt{570}$
41.8 ± 1.8
$40.0 < \mu < 43.6$
b. $E = .5$
$\alpha = .01$
$n = [z_{.005} \cdot \sigma / E]^2$
$= [2.757 \cdot 16.7 / .5]^2$
$= 7396.86$ rounded up to 7397

23. $n = 16$
$\Sigma x = 133.3$
$\Sigma x^2 = 1167.99$
$\overline{x} = 8.33$
$s = 1.957$ [do not round, store all the digits in the calculator]
$n \leq 30$ and σ unknown, use t
$\overline{x} \pm t_{15,.025} \cdot \sigma/\sqrt{n}$
$8.33 \pm 2.132 \cdot 1.957/\sqrt{16}$
8.33 ± 1.04
$7.29 < \mu < 9.37$

25. $n = 62$
$\Sigma x = 584.54$
$\Sigma x^2 = 6570.8216$
$\overline{x} = 9.428$
$s = 4.168$ [do not round, store all the digits in the calculator]
$n > 30$, use z (with s for σ)
$\overline{x} \pm z_{.025} \cdot \sigma/\sqrt{n}$
$9.428 \pm 1.960 \cdot 4.168/\sqrt{62}$
9.428 ± 1.038
$8.390 < \mu < 10.466$

27. a. Use only films with the R rating.
   $n = 35$
   $\Sigma x = 3890$
   $\Sigma x^2 = 446620$
   $\overline{x} = 111.1$
   $s = 20.49$ [do not round, store all the digits in the calculator]
   $n > 30$, use z (with s for σ)
   $\overline{x} \pm z_{.025} \cdot \sigma/\sqrt{n}$
   $111.1 \pm 1.960 \cdot 20.49/\sqrt{35}$
   111.1 ± 6.8
   $104.3 < \mu < 117.9$

   b. Use only films with the PG or PG-13 rating.
   $n = 23$
   $\Sigma x = 2552$
   $\Sigma x^2 = 297902$
   $\overline{x} = 111.0$
   $s = 25.89$ [do not round, store all the digits in the calculator]
   $n \leq 30$ and σ unknown, use t
   $\overline{x} \pm t_{22,.025} \cdot \sigma/\sqrt{n}$
   $111.0 \pm 2.074 \cdot 25.89/\sqrt{23}$
   111.0 ± 11.2
   $99.8 < \mu < 122.2$

   c. Since there were 35 (i.e., more than 30) films with the R rating, part (a) used z scores; since there were only 23 (i.e., less than or equal to 30) films with the PG or PG-13 rating, part (b) used t scores. The confidence interval in part (b) is wider because (1) the PG and PG-13 films exhibited more variability (as measured by the standard deviation) than the R films and (2) the larger number of R films produced a larger denominator in the standard error and a smaller table value (i.e., t or z value) by which to multiply the standard error to determine E.

29. $n > 30$, use z (with s for σ)

$\quad$ NOTE: For the original 23 data values

$E = z_{.025} \cdot \sigma / \sqrt{n}$

$\quad = 1.960 \cdot 7.26 / \sqrt{850}$

$\quad = .5$

$\Sigma x = 19962.2$

$\Sigma x^2 = 13284510.72$

$\bar{x} = 665.41$

$s = 7.26$

31. $E = 1.5$

$\quad \alpha = .02$

$\quad N = 200$

$\quad n = [N\sigma^2(z_{.01})^2]/[(N-1)E^2 + \sigma^2(z_{.01})^2]$

$\quad = [(200)(15)^2(2.327)^2]/[(199)(1.5)^2 + (15)^2(2.327)^2]$

$\quad = [243671.8]/[1666.109]$

$\quad = 146.25$ rounded up to 147

6-3 Estimating a Population Proportion

IMPORTANT NOTE: When calculating confidence intervals using the formula

$\quad \hat{p} \pm E$

$\quad \hat{p} \pm z_{\alpha/2}\sqrt{\hat{p}\hat{q}/n}$

do not round off in the middle of the problem. This may be accomplished conveniently on most calculators having a memory as follows.

(1) Calculate $\hat{p} = x/n$ and STORE the value

(2) Calculate E as 1 - RECALL = * RECALL = ÷ n = √ * $z_{\alpha/2}$ =

(3) With the value of E showing on the display, the upper confidence limit is calculated by + RECALL.

(4) With the value of the upper confidence limit showing on the display, the lower confidence limit is calculated by - RECALL ± + RECALL

You must become familiar with your own calculator. [Do your homework using the same type of calculator you will be using for the exams.] The above procedure works on most calculators; make certain you understand why it works and verify whether it works on your calculator. If it does not seem to work on your calculator, or if your calculator has more than one memory so that you can STORE both $\hat{p}$ and E at the same time, ask your instructor for assistance.

NOTE: It should be true that $0 \leq \hat{p} \leq 1$ and that $E \leq .5$ [usually, <u>much</u> less than .5]. If such is not the case, an error has been made.

1. $\hat{p} = x/n$

$\quad = 100/500$

$\quad = .20$

$\quad \alpha = .05$

$\quad E = z_{.025}\sqrt{\hat{p}\hat{q}/n}$

$\quad = 1.960\sqrt{(.20)(.80)/500}$

$\quad = .0351$

3. $\hat{p} = x/n$

$\quad = 325/1068$

$\quad = .304$

$\quad \alpha = .05$

$\quad E = z_{.025}\sqrt{\hat{p}\hat{q}/n}$

$\quad = 1.960\sqrt{(.304)(.696)/1068}$

$\quad = .0276$

5. $\hat{p} = x/n$
   $= 100/400$
   $= .250$
   $\alpha = .05$
   $\hat{p} \pm z_{.025}\sqrt{\hat{p}\hat{q}/n}$
   $.250 \pm 1.960\sqrt{(.250)(.750)/400}$
   $.250 \pm .042$
   $.208 < p < .292$

7. $\hat{p} = x/n$
   $= 309/512$
   $= .604$
   $\alpha = .02$
   $\hat{p} \pm z_{.01}\sqrt{\hat{p}\hat{q}/n}$
   $.604 \pm 2.327\sqrt{(.604)(.396)/512}$
   $.604 \pm .050$
   $.553 < p < .654$

9. $E = .02$
   $\alpha = .03$
   $\hat{p}$ unknown, use $\hat{p} = .5$
   $n = [(z_{.015})^2\hat{p}\hat{q}]/E^2$
   $= [(2.17)^2(.5)(.5)]/(.02)^2$
   $= 2943.06$ rounded up to 2944

   NOTE: To find $z_{.015}$ in Table A-3,
   find $A = .4850$ (or closest entry)
   read z from the margins of the table

11. $\hat{p} = .63$
    $\alpha = .02$
    $\hat{p} \pm z_{.01}\sqrt{\hat{p}\hat{q}/n}$
    $.63 \pm 2.327\sqrt{(.63)(.37)/1500}$
    $.63 \pm .03$
    $.60 < p < .66$
    NOTE: Since $\hat{p}$ was given with only two decimal accuracy and the actual value of x was not given, the final answer is limited to two decimal accuracy. Any x from 938 to 952 rounds to 63% with varying digits for the next decimal point.

13. $\hat{p} = x/n$
    $= 1280/2000$
    $= .640$
    $\alpha = .05$
    $\hat{p} \pm z_{.025}\sqrt{\hat{p}\hat{q}/n}$
    $.640 \pm 1.960\sqrt{(.640)(.360)/2000}$
    $.640 \pm .021$
    $.553 < p < .654$

15. $E = .02$
    $\alpha = .10$
    $\hat{p} = .29$
    $n = [(z_{.05})^2\hat{p}\hat{q}]/E^2$
    $= [(1.645)^2(.29)(.71)]/(.02)^2$
    $= 1392.93$ rounded up to 1393

17. $\hat{p} = .75$
$\alpha = .10$
$\hat{p} \pm z_{.05}\sqrt{\hat{p}\hat{q}/n}$
$.75 \pm 1.645\sqrt{(.75)(.25)/600}$
$.75 \pm .03$
$.72 < p < .78$
NOTE: Since $\hat{p}$ was given with only two decimal accuracy and the actual value of x was not given, the final answer is limited to two decimal accuracy. Any x from 447 to 453 rounds to 75% with varying digits for the next decimal point.

19. $\hat{p} = x/n$
$= 312/650$
$= .480$
$\alpha = .05$
$\hat{p} \pm z_{.025}\sqrt{\hat{p}\hat{q}/n}$
$.480 \pm 1.960\sqrt{(.480)(.520)/650}$
$.480 \pm .038$
$.442 < p < .518$

21. $E = .03$
$\alpha = .06$
$\hat{p} = .27$
$n = [(z_{.03})^2\hat{p}\hat{q}]/E^2$
$= [(1.88)^2(.27)(.73)]/(.03)^2$
$= 774.03$ rounded up to 775

NOTE: To find $z_{.03}$ in Table A-3,
find $A = .4700$ (or closest entry)
read z from the margins of the table

23. $\hat{p} = x/n$
$= 180/650$
$= .277$
$\alpha = .05$
$\hat{p} \pm z_{.025}\sqrt{\hat{p}\hat{q}/n}$
$.277 \pm 1.960\sqrt{(.277)(.723)/650}$
$.277 \pm .034$
$.243 < p < .311$

25. $\hat{p} = .72$
$\alpha = .01$
$\hat{p} \pm z_{.005}\sqrt{\hat{p}\hat{q}/n}$
$.72 \pm 2.575\sqrt{(.72)(.28)/4664}$
$.72 \pm .02$
$.70 < p < .74$
NOTE: Since $\hat{p}$ was given with only two decimal accuracy and the actual value of x was not given, the final answer is limited to two decimal accuracy. Any x from 3335 to 3381 rounds to 72% with varying digits for the next decimal point.

27. There are 36 of the 125 individuals that studied beyond high school.
$\hat{p} = x/n$
$= 36/125$
$= .288$
$\alpha = .05$
$\hat{p} \pm z_{.025}\sqrt{\hat{p}\hat{q}/n}$
$.288 \pm 1.960\sqrt{(.288)(.712)/125}$
$.288 \pm .079$
$.209 < p < .367$

29. There are 35 of the 60 movies that received the R rating.

$\hat{p} = x/n$

$\quad = 35/60$

$\quad = .583$

$\alpha = .05$

$\hat{p} \pm z_{.025}\sqrt{\hat{p}\hat{q}/n}$

$.583 \pm 1.960\sqrt{(.583)(.417)/60}$

$.583 \pm .125$

$.459 < p < .708$

31. $E = .01$

$\alpha = .05$ [since 19/20 implies 95% confidence]

$\hat{p}$ unknown, use $\hat{p} = .5$

$n = [(z_{.025})^2\hat{p}\hat{q}]/E^2$

$\quad = [(1.960)^2(.5)(.5)]/(.01)^2$

$\quad = 9604$

33. $E = .002$

$\hat{p} = .08$

$n = 47000$

Solve $E = z_{\alpha/2}\sqrt{\hat{p}\hat{q}/n}$ for $z_{\alpha/2}$ to get

$z_{\alpha/2} = E/\sqrt{\hat{p}\hat{q}/n}$

$\quad = .002/\sqrt{(.08)(.92)/47000}$

$\quad = .002/.00125$

$\quad = 1.60$

Since $P(-1.60 < z < 1.60) = .4452 + .4452$

$\qquad\qquad\qquad\qquad\qquad = .8904,$

the level of confidence is .8904, or about 89%.

35. a.

$$E = z_{\alpha/2}\sqrt{\hat{p}\hat{q}/n}\sqrt{(N-n)/(N-1)}$$

$E^2 = (z_{\alpha/2})^2 \cdot [\hat{p}\hat{q}/n] \cdot [(N-n)/(N-1)]$ squaring both sides

$n(N-1)E^2 = (z_{\alpha/2})^2\hat{p}\hat{q}(N-n)$ multiplying by n(N-1)

$n(N-1)E^2 = (z_{\alpha/2})^2\hat{p}\hat{q}N - (z_{\alpha/2})^2\hat{p}\hat{q}n$ distributing $(z_{\alpha/2})^2\hat{p}\hat{q}$

$(z_{\alpha/2})^2\hat{p}\hat{q}n + n(N-1)E^2 = (z_{\alpha/2})^2\hat{p}\hat{q}N$ adding $(z_{\alpha/2})^2\hat{p}\hat{q}$

$n[(z_{\alpha/2})^2\hat{p}\hat{q} + (N-1)E^2] = (z_{\alpha/2})^2\hat{p}\hat{q}N$ factoring out n

$n = (z_{\alpha/2})^2\hat{p}\hat{q}N/[(z_{\alpha/2})^2\hat{p}\hat{q} + (N-1)E^2]$

b. $E = .03$

$\alpha = .06$

$\hat{p} = .27$

$n = (z_{\alpha/2})^2\hat{p}\hat{q}N/[(z_{\alpha/2})^2\hat{p}\hat{q} + (N-1)E^2]$

$\quad = (1.88)^2(.27)(.73)(500)/[(1.88)^2(.27)(.73) + (499)(.03)^2]$

$\quad = 348.315/1.14573$

$\quad = 304.01$ rounded up to 305

6-4   Estimating a Population Variance

1. a. From the 95% and σ^2 section of Table 6-2, n=97.

b. The best point estimate for σ^2 is $s^2 = 144.0$.

c. $\chi^2_L = \chi^2_{26,.975} = 13.844$

$\chi^2_R = \chi^2_{26,.025} = 41.923$

d. $\qquad (n-1)s^2/\chi^2_R < \sigma^2 < (n-1)s^2/\chi^2_L$

$(26)(144.0)/41.923 < \sigma^2 < (26)(144.0)/13.844$

$\qquad\qquad 89.3 < \sigma^2 < 270.4$

e. Taking the square roots, $9.5 < \sigma < 16.4$

3. summary information
   $n = 10$
   $\Sigma x = 1018$
   $\Sigma x^2 = 104908$
   $\bar{x} = 101.8$
   $s^2 = 141.7$
   a. $s^2 = 141.7$
   b. $(n-1)s^2/\chi^2_{9,.025} < \sigma^2 < (n-1)s^2/\chi^2_{9,.975}$
      $(9)(141.7)/19.023 < \sigma^2 < (9)(141.7)/2.700$
             $67.1 < \sigma^2 < 472.4$
              $8.2 < \sigma < 21.7$
   c. Yes, $8.2 < 15 < 21.7$

5. $(n-1)s^2/\chi^2_{100,.005} < \sigma^2 < (n-1)s^2/\chi^2_{100,.995}$
   $(100)(1.68)^2/140.169 < \sigma^2 < (100)(1.68)^2/67.328$
             $2.01 < \sigma^2 < 4.19$
             $1.42 < \sigma < 2.05$

7. $(n-1)s^2/\chi^2_{19,.025} < \sigma^2 < (n-1)s^2/\chi^2_{19,.975}$
   $(19)(1.30)^2/32.852 < \sigma^2 < (19)(1.30)^2/8.907$
             $.98 < \sigma^2 < 3.61$
             $.99 < \sigma < 1.90$

9. $(n-1)s^2/\chi^2_{49,.005} < \sigma^2 < (n-1)s^2/\chi^2_{49,.995}$
   $(49)(.068)^2/79.490 < \sigma^2 < (49)(.068)^2/27.991$
             $.003 < \sigma^2 < .008$

11. $(n-1)s^2/\chi^2_{100,.01} < \sigma^2 < (n-1)s^2/\chi^2_{100,.99}$
   $(100)(41.0)^2/135.807 < \sigma^2 < (100)(41.0)^2/70.065$
             $1237.8 < \sigma^2 < 2399.2$
             $35.2 < \sigma < 49.0$

13. summary information
   $n = 16$
   $\Sigma x = 133.3$
   $\Sigma x^2 = 1167.99$
   $\bar{x} = 8.33$
   $s^2 = 3.829$
        $(n-1)s^2/\chi^2_{15,.025} < \sigma^2 < (n-1)s^2/\chi^2_{15,.975}$
   $(15)(3.829)/27.488 < \sigma^2 < (15)(3.829)/6.262$
             $2.09 < \sigma^2 < 9.17$
             $1.45 < \sigma < 3.03$

15. summary information
   $n = 62$
   $\Sigma x = 584.54$
   $\Sigma x^2 = 6570.8216$
   $\bar{x} = 9.428$
   $s^2 = 17.373$
   $(n-1)s^2/\chi^2_{61,.025} < \sigma^2 < (n-1)s^2/\chi^2_{61,.975}$
   $(61)(17.373)/83.298 < \sigma^2 < (61)(17.373)/40.482$
   $12.722 < \sigma^2 < 26.178$
   $3.567 < \sigma < 5.116$

17. The given interval $2.8 < \sigma < 6.0$
$$7.84 < \sigma^2 < 36.00$$
and the usual calculations $(n-1)s^2/\chi^2_{19,\alpha/2} < \sigma^2 < (n-1)s^2/\chi^2_{19,1-\alpha/2}$
$$(19)(3.8)^2/\chi^2_{19,\alpha/2} < \sigma^2 < (19)(3.8)^2/\chi^2_{19,1-\alpha/2}$$
$$274.36/\chi^2_{19,\alpha/2} < \sigma^2 < 274.36/\chi^2_{19,1-\alpha/2}$$

imply that $7.84 = 274.37/\chi^2_{19,\alpha/2}$ and $36.00 = 274.36/\chi^2_{19,1-\alpha/2}$

$\chi^2_{19,\alpha/2} = 274.36/7.84$ $\chi^2_{19,1-\alpha/2} = 274.36/36.00$
$= 34.99$ $= 7.62$

The closest entries in Table A-4 are $\chi^2_{19,\alpha/2} = 34.805$ and $\chi^2_{19,1-\alpha/2} = 7.633$
which imply $\alpha/2 = .01$ $1 - \alpha/2 = .99$
$\alpha = .01$ $\alpha/2 = .01$
$\alpha = .02$

The level of confidence is therefore is $1-\alpha = 98\%$.

19. $\chi^2 = \frac{1}{2}[\pm z_{.025} + \sqrt{2 \cdot df - 1}]^2$
$ = \frac{1}{2}[\pm 1.960 + \sqrt{2 \cdot (771) - 1}]^2$
$ = \frac{1}{2}[\pm 1.960 + 39.256]^2$
$ = \frac{1}{2}[37.296]^2$ or $\frac{1}{2}[41.216]^2$
$ = 695.48$ or 849.36

$$(n-1)s^2/\chi^2_{771,.025} < \sigma^2 < (n-1)s^2/\chi^2_{771,.975}$$
$$(771)(2.8)^2/849.36 < \sigma^2 < (771)(2.8)^2/695.48$$
$$7.117 < \sigma^2 < 8.691$$
$$2.7 < \sigma < 2.9$$

Review Exercises

1. a. The best point estimate for μ is $\overline{x} = 17.6$
   b. $n > 30$, use z (with s for σ)
   $\overline{x} \pm z_{.025} \cdot \sigma/\sqrt{n}$
   $17.6 \pm 1.960 \cdot 9.3/\sqrt{50}$
   17.6 ± 2.6
   $15.0 < \mu < 20.2$

3. $(n-1)s^2/\chi^2_{24,.025} < \sigma^2 < (n-1)s^2/\chi^2_{24,.975}$
   $(24)(3.74)^2/39.364 < \sigma^2 < (24)(3.74)^2/12.401$
   $8.528 < \sigma^2 < 27.071$
   $2.92 < \sigma < 5.20$

5. NOTE: $n = 70 + 711 = 781$
   $\hat{p} = x/n$
   $\phantom{\hat{p}} = 70/781$
   $\phantom{\hat{p}} = .0896$
   $\alpha = .05$
   $\hat{p} \pm z_{.025}\sqrt{\hat{p}\hat{q}/n}$
   $.0896 \pm 1.960\sqrt{(.0896)(.9104)/781}$
   $.0896 \pm .0200$
   $.0696 < p < .1097$

7. $E = 250$
   $\alpha = .04$
   $n = [z_{.02} \cdot \sigma/E]^2$
   $\quad = [2.05 \cdot 3050/250]^2$
   $\quad = 625.50$ rounded up to 626

9. $\underline{n} > 30$, use $\underline{z}$ (with s for σ)
   $\bar{x} \pm z_{.005} \cdot \sigma/\sqrt{n}$
   $40.7 \pm 2.575 \cdot 10.2/\sqrt{40}$
   40.7 ± 4.2
   $36.5 < \mu < 44.9$

11. $E = .04$
    $\alpha = .10$
    $\hat{p}$ unknown, use $\hat{p} = .5$
    $n = [(z_{.05})^2 \hat{p}\hat{q}]/E^2$
    $\quad = [(1.645)^2(.5)(.5)]/(.04)^2$
    $\quad = 422.82$ rounded up to 423

13. $E = 4$
    $\alpha = .05$
    $n = [z_{.025} \cdot \sigma/E]^2$
    $\quad = [1.960 \cdot 41.0/4]^2$
    $\quad = 403.61$ rounded up to 404

15. $\hat{p} = .24$
    $\alpha = .01$
    $\hat{p} \pm z_{.005}\sqrt{\hat{p}\hat{q}/n}$
    $.24 \pm 2.575\sqrt{(.24)(.76)/1998}$
    $.24 \pm .02$
    $.22 < p < .26$
    NOTE: Since $\hat{p}$ was given with only two decimal accuracy and the actual value of x was not given, the final answer is limited to two decimal accuracy. Any x from 470 to 489 rounds to 24% with varying digits for the next decimal point.

17. $n = 16$
    $\Sigma x = 1162.6$
    $\Sigma x^2 = 84577.34$
    $\bar{x} = 72.66$ [do not round, store <u>all</u> the digits in the calculator]
    $s = 2.581$ [do not round, store <u>all</u> the digits in the calculator]
    $\underline{n} \leq 30$ and σ unknown, use t
    $\bar{x} \pm t_{15,.025} \cdot \sigma/\sqrt{n}$
    $72.66 \pm 2.132 \cdot 2.581/\sqrt{16}$
    72.66 ± 1.38
    $71.29 < \mu < 74.04$

Chapter 7

Hypothesis Testing

7-2 Fundamentals of Hypothesis testing

1. a. $\mu = 120$
   b. $H_o: \mu = 120$
   c. $H_1: \mu \neq 120$
   d. two-tailed
   e. rejecting the hypothesis that the mean IQ is 120 when it really is 120
   f. failing to reject the hypothesis that the mean IQ is 120 when it really is not 120
   g. there is sufficient evidence to reject the claim that the mean IQ is 120
   h. there is not sufficient evidence to reject the claim that the mean IQ is 120

3. a. $\mu > 5$
   b. $H_o: \mu \leq 5$
   c. $H_1: \mu > 5$
   d. right-tailed
   e. rejecting the hypothesis that the mean time is less than or equal to 5 years when it really is
   f. failing to reject the hypothesis that the mean time is less than or equal to 5 years when it really is not
   g. there is sufficient evidence to support the claim that the mean time is greater than 5 years
   h. there is not sufficient evidence to support the claim that the mean time is greater than 5 years

5. a. $\mu \geq 10$
   b. $H_o: \mu \geq 10$
   c. $H_1: \mu < 10$
   d. left-tailed
   e. rejecting the hypothesis that the mean age is at least 10 years when it really is
   f. failing to reject the hypothesis that the mean age is at least 10 years when it really is not
   g. there is sufficient evidence to reject the claim that the mean age is at least 10 years
   h. there is not sufficient evidence to reject the claim that the mean age is at least 10 years

7. the critical z values are $\pm z_{.005} = \pm 2.575$
   the critical region is z < -2.575
   $\quad\quad\quad\quad\quad\quad$ z > 2.575

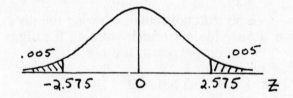

9. the critical z value is $z_{.01} = 2.327$ [from the last row of Table A-3]
   the critical region is z > 2.327

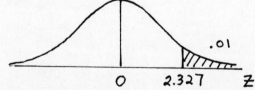

11. the critical z value is $-z_{.05} = -1.645$
    the critical region is $z < -1.645$

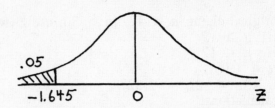

13. a. The conclusion indicates H_o was rejected; the null hypothesis must have been $\mu \leq 30$.
    b. The conclusion indicated H_o was not rejected; the null hypothesis must have been $\mu \geq 5$.
    c. The conclusion indicates H_o was rejected; the null hypothesis must have been $\mu = 70$.

15. Mathematically, in order for α to equal 0 the magnitude of the critical value would have to be infinite. Practically, the only way never to make a type I error is to always fail to reject H_o. From either perspective, the only way to achieve $\alpha = 0$ is to never reject H_o no matter how extreme the sample data might be.

7-3 Testing a Claim about a Mean: Large Samples

1. original claim: $\mu = 100$ [$n > 30$, use z (with s for σ)]
   H_o: $\mu = 100$
   H_1: $\mu \neq 100$
   $\alpha = .01$
   C.R. $z < -z_{.005} = -2.575$
        $z > z_{.005} = 2.575$
   calculations:
   $$z_{\bar{x}} = (\bar{x} - \mu)/\sigma_{\bar{x}}$$
   $$= (100.8 - 100)/(5/\sqrt{81})$$
   $$= .8/.556$$
   $$= 1.44$$
   conclusion:
   Do not reject H_o; there is not sufficient evidence to conclude that $\mu \neq 100$.

3. original claim: $\mu = 500$ [$n > 30$, use z (with s for σ)]
   H_o: $\mu = 500$
   H_1: $\mu \neq 500$
   $\alpha = .10$
   C.R. $z < -z_{.05} = -1.645$
        $z > z_{.05} = 1.645$
   calculations:
   $$z_{\bar{x}} = (\bar{x} - \mu)/\sigma_{\bar{x}}$$
   $$= (510 - 500)/(50/\sqrt{300})$$
   $$= 10/2.887$$
   $$= 3.464$$
   conclusion:
   Reject H_o; there is sufficient evidence to conclude that $\mu \neq 500$ (in fact, $\mu > 500$).

5. original claim: $\mu < 12$ [n > 30, use z (with s for σ)]
   H_o: $\mu \geq 12$
   H_1: $\mu < 12$
   $\alpha = .01$
   C.R. $z < -z_{.01} = -2.327$
   calculations:
   $z_{\bar{x}} = (\bar{x} - \mu)/\sigma_{\bar{x}}$
   $= (11.82 - 12)/(.38/\sqrt{36})$
   $= -.18/.0633$
   $= -2.842$
   conclusion:
   Reject H_o; there is sufficient evidence to conclude that $\mu < 12$.

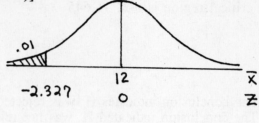

7. original claim: $\mu > .21$ [n > 30, use z (with s for σ)]
   H_o: $\mu \leq .21$
   H_1: $\mu > .21$
   $\alpha = .01$
   C.R. $z > z_{.01} = 2.327$
   calculations:
   $z_{\bar{x}} = (\bar{x} - \mu)/\sigma_{\bar{x}}$
   $= (.83 - .21)/(.24/\sqrt{32})$
   $= .62/.0424$
   $= 14.614$
   conclusion:
   Reject H_o; there is sufficient evidence to conclude that $\mu > .21$.

9. original claim: $\mu < 1.39$ [n > 30, use z (with s for σ)]
   H_o: $\mu \geq 1.39$
   H_1: $\mu < 1.39$
   $\alpha = .01$
   C.R. $z < -z_{.01} = -2.327$
   calculations:
   $z_{\bar{x}} = (\bar{x} - \mu)/\sigma_{\bar{x}}$
   $= (.83 - 1.39)/(.16/\sqrt{123})$
   $= -.56/.0144$
   $= -38.817$
   conclusion:
   Reject H_o; there is sufficient evidence to conclude that $\mu < 1.39$.

11. original claim: $\mu > 0$ [n > 30, use z (with s for σ)]
   H_o: $\mu \leq 0$
   H_1: $\mu > 0$
   $\alpha = .05$
   C.R. $z > z_{.05} = 1.645$
   calculations:
   $z_{\bar{x}} = (\bar{x} - \mu)/\sigma_{\bar{x}}$
   $= (.6 - 0)/(3.8/\sqrt{75})$
   $= .6/.439$
   $= 1.367$
   conclusion:
   Do not reject H_o; there is not sufficient evidence to conclude that $\mu > 0$.

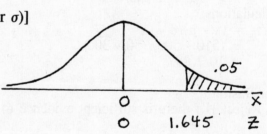

13. original claim: $\mu > 51.0$ [n > 30, use z (with s for σ)]
   H_o: $\mu \leq 51.0$
   H_1: $\mu > 51.0$
   $\alpha = .05$
   C.R. z > $z_{.05}$ = 1.645
   calculations:
   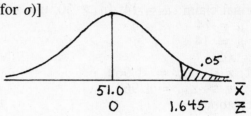

   $\quad z_{\bar{x}} = (\bar{x} - \mu)/\sigma_{\bar{x}}$
   $\quad\quad = (52.4 - 51.0)/(13.14/\sqrt{150})$
   $\quad\quad = 1.4/1.073$
   $\quad\quad = 1.305$
   conclusion:
   $\quad$ Do not reject H_o; there is not sufficient evidence to conclude that $\mu > 51.0$.

15. original claim: $\mu < 24$ [n > 30, use z (with s for σ)]
   H_o: $\mu \geq 24$
   H_1: $\mu < 24$
   $\alpha = .05$
   C.R. z < $-z_{.05}$ = -1.645
   calculations:
   $\quad z_{\bar{x}} = (\bar{x} - \mu)/\sigma_{\bar{x}}$
   $\quad\quad = (22.1 - 24)/(8.6/\sqrt{200})$
   $\quad\quad = -1.9/.608$
   $\quad\quad = -3.124$
   conclusion:
   $\quad$ Reject H_o; there is sufficient evidence to conclude that $\mu < 24$.

17. original claim: $\mu = 7124$ [n > 30, use z (with s for σ)]
   H_o: $\mu \geq 7124$
   H_1: $\mu < 7124$
   $\alpha = .05$
   C.R. z < $-z_{.05}$ = -1.645
   calculations:
   $\quad z_{\bar{x}} = (\bar{x} - \mu)/\sigma_{\bar{x}}$
   $\quad\quad = (6047 - 7124)/(2944/\sqrt{750})$
   $\quad\quad = -1077/107.500$
   $\quad\quad = -10.019$
   conclusion:
   $\quad$ Reject H_o; there is sufficient evidence to conclude that $\mu < 7124$.

19. original claim: $\mu > 40,000$ [n > 30, use z (with s for σ)]
   H_o: $\mu \leq 40,000$
   H_1: $\mu > 40,000$
   $\alpha = .05$
   C.R. z > $z_{.05}$ = 1.645
   calculations:
   $\quad z_{\bar{x}} = (\bar{x} - \mu)/\sigma_{\bar{x}}$
   $\quad\quad = (41.182 - 40,000)/(19,990/\sqrt{1700})$
   $\quad\quad = 1182/484.829$
   $\quad\quad = 2.438$
   conclusion:
   $\quad$ Reject H_o; there is sufficient evidence to conclude that $\mu > 40,000$.

21. original claim: $\mu = 14$ $[n > 30$, use z (with s for σ)]
    $H_o: \mu = 14$
    $H_1: \mu \neq 14$
    $\alpha = .05$
    C.R. $z < -z_{.025} = -1.960$
    $\quad\quad z > z_{.025} = 1.960$
    calculations:
    $\quad z_{\bar{x}} = (\bar{x} - \mu)/\sigma_{\bar{x}}$
    $\quad\quad = (13.41 - 14)/(8.28/\sqrt{40})$
    $\quad\quad = -.59/1.309$
    $\quad\quad = -.451$
    conclusion:
    $\quad$ Do not reject H_o; there is not sufficient evidence to conclude that $\mu \neq 14$.

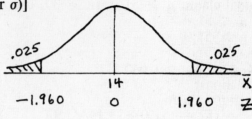

23. original claim: $\mu = 5.670$ $[n > 30$, use z (with s for σ)]
    $H_o: \mu = 5.670$
    $H_1: \mu \neq 5.670$
    $\alpha = .01$
    C.R. $z < -z_{.005} = -2.575$
    $\quad\quad z > z_{.005} = 2.575$
    calculations:
    $\quad z_{\bar{x}} = (\bar{x} - \mu)/\sigma_{\bar{x}}$
    $\quad\quad = (6.622 - 5.670)/(.068/\sqrt{50})$
    $\quad\quad = -.048/.00962$
    $\quad\quad = -4.991$
    conclusion:
    $\quad$ Reject H_o; there is sufficient evidence to conclude that $\mu \neq 5.670$ (in fact, $\mu < 5.670$).
    The fact that the mean weight of worn quarters found in circulation is less than 5.670 is not
    $\quad$ evidence that the mean weight of uncirculated quarters straight from the mint is different
    $\quad$ from 5.670.

25. original claim: $\mu = 105$ $[n > 30$, use z (with s for σ)]
    NOTE: From the boxplot, $\bar{x} = 107$ and $P_{75} = 115$.
    $\quad\quad$ Assuming a normal distribution, P_{75} corresponds to z = .67.
    $\quad\quad$ Solving for s, $z = (x - \bar{x})/s$
    $\quad\quad\quad\quad\quad\quad .67 = (115 - 107)/s$
    $\quad\quad\quad\quad\quad\quad .67 = 8/s$
    $\quad\quad\quad\quad\quad\quad\quad s = 8/.67$
    $\quad\quad\quad\quad\quad\quad\quad\quad = 11.940$
    $H_o: \mu = 105$
    $H_1: \mu \neq 105$
    $\alpha = .03$
    C.R. P-value $< .03$
    calculations:
    $\quad z_{\bar{x}} = (\bar{x} - \mu)/\sigma_{\bar{x}}$
    $\quad\quad = (107 - 105)/(11.940/\sqrt{93})$
    $\quad\quad = 2/1.238$
    $\quad\quad = 1.62$
    P-value $= 2 \cdot P(z > 1.62)$
    $\quad\quad\quad\quad = 2 \cdot [.5000 - .4774]$
    $\quad\quad\quad\quad = 2 \cdot [.0526]$
    $\quad\quad\quad\quad = .1052$
    conclusion:
    $\quad$ Do not reject H_o; there is not sufficient evidence to conclude that $\mu \neq 105$.

27. original claim: $\mu = 23{,}460$ [n > 30, use z (with s for σ)]
   H_o: $\mu = 23{,}460$
   H_1: $\mu \neq 23{,}460$
   $\alpha = .02$
   C.R. $z < -z_{.01} = -2.327$
   $\qquad z > z_{.01} = 2.327$
   calculations:

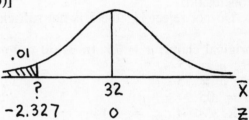

$$z_{\bar{x}} = (\bar{x} - \mu)/\sigma_{\bar{x}}$$
$$2.327 = (\bar{x} - 23{,}460)/(3750/\sqrt{50})$$
$$2.327 = (\bar{x} - 23{,}460)/530.33$$
$$1234.08 = \bar{x} - 23{,}460$$
$$24{,}694.08 = \bar{x}$$

NOTE: There is no conclusion because this exercise is not asking for the completion of a test of hypotheses. In fact, the sample mean is irrelevant to the question posed. The exercise asks for the $\bar{x}$ that corresponds to the z score that marks the beginning of the upper portion of the critical region.

29. original claim: $\mu < 32$ [n > 30, use z (with s for σ)]
   H_o: $\mu \geq 32$
   H_1: $\mu < 32$
   $\alpha = .01$
   C.R. $z < -z_{.01} = -2.327$
   calculations:

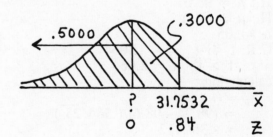

$$z_{\bar{x}} = (\bar{x} - \mu)/\sigma_{\bar{x}}$$
$$-2.327 = (\bar{x} - 32)/(.75/\sqrt{50})$$
$$-2.327 = (\bar{x} - 32)/.106$$
$$-.2468 = \bar{x} - 32$$
$$31.7532 = \bar{x}$$

NOTE: There is no conclusion because this exercise is not asking for the completion of a test of hypothesis. In fact, the sample mean is irrelevant to the question posed. The above calculations find the $\bar{x}$ that corresponds to the z score that marks the beginning of the critical region. In terms of $\bar{x}$, then, the critical region is $\bar{x} < 31.7532$.

The exercise asks for the value of μ for which P(rejecting H_o) = P($\bar{x} < 31.7532$) = .80.
NOTE: The z score with .80 below it [i.e., with A = .3000 in Table A-1] is z = .84.

$$z_{\bar{x}} = (\bar{x} - \mu)/\sigma_{\bar{x}}$$
$$.84 = (31.7532 - \mu)/(.75/\sqrt{50})$$
$$.84 = (31.7532 - \mu)/.106$$
$$.0891 = 31.7532 - \mu$$
$$\mu = 31.7532 - .0891$$
$$= 31.6641$$

7-4 Testing a Claim about a Mean: Small Samples

1. a. $\pm t_{26,.025} = \pm 2.056$
   b. $t_{16,.10} = 1.337$
   c. $-t_{5,.01}$ -3.365

3. original claim: $\mu \leq 10$ [n ≤ 30 and σ unknown, use t]
   H_o: $\mu \leq 10$
   H_1: $\mu > 10$
   $\alpha = .05$
   C.R. $t > t_{8,.05} = 1.860$
   calculations:
   $t_{\bar{x}} = (\bar{x} - \mu)/s_{\bar{x}}$
   $= (11 - 10)/(2/\sqrt{9})$
   $= 1/.667$
   $= 1.500$
   conclusion:
   Do not reject H_o; there is not sufficient evidence to conclude that $\mu > 10$.

5. original claim: $\mu = 75$ [n ≤ 30 and σ unknown, use t]
   H_o: $\mu = 75$
   H_1: $\mu \neq 75$
   $\alpha = .05$
   C.R. $t < -t_{14,.025} = -2.145$
   $t > t_{14,.025} = 2.145$
   calculations:
   $t_{\bar{x}} = (\bar{x} - \mu)/s_{\bar{x}}$
   $= (77.6 - 75)/(5/\sqrt{15})$
   $= 2.6/1.291$
   $= 2.014$
   conclusion:
   Do not reject H_o; there is not sufficient evidence to conclude that $\mu \neq 75$.

7. original claim: $\mu = 98.6$ [n ≤ 30 and σ unknown, use t]
   H_o: $\mu = 98.6$
   H_1: $\mu \neq 98.6$
   $\alpha = .05$
   C.R. $t < -t_{24,.025} = -2.064$
   $t > t_{24,.025} = 2.064$
   calculations:
   $t_{\bar{x}} = (\bar{x} - \mu)/s_{\bar{x}}$
   $= (98.24 - 98.6)/(.56/\sqrt{25})$
   $= -.36/.112$
   $= -3.214$
   conclusion:
   Reject H_o; there is sufficient evidence to conclude that $\mu \neq 98.6$ (in fact, $\mu < 98.6$).

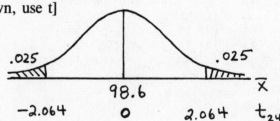

9. original claim: $\mu > 32$ [n ≤ 30 and σ unknown, use t]
   H_o: $\mu \leq 32$
   H_1: $\mu > 32$
   $\alpha = .10$
   C.R. t $> t_{26,.10} = 1.315$
   calculations:
   $\quad t_{\bar{x}} = (\bar{x} - \mu)/s_{\bar{x}}$
   $\quad\quad = (32.2 - 32)/(.4/\sqrt{27}\,)$
   $\quad\quad = .2/.0770$
   $\quad\quad = 2.598$
   conclusion:
   $\quad$ Reject H_o; there is sufficient evidence to conclude that $\mu > 32$.

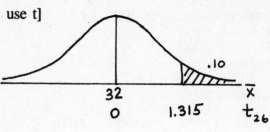

11. original claim: $\mu > 1800$ [n ≤ 30 and σ unknown, use t]
   H_o: $\mu \leq 1800$
   H_1: $\mu > 1800$
   $\alpha = .01$
   C.R. t $> t_{11,.01} = 2.718$
   calculations:
   $\quad t_{\bar{x}} = (\bar{x} - \mu)/s_{\bar{x}}$
   $\quad\quad = (2133 - 1800)/(345/\sqrt{12}\,)$
   $\quad\quad = 333/99.593$
   $\quad\quad = 3.344$
   conclusion:
   $\quad$ Reject H_o; there is sufficient evidence to conclude that $\mu > 1800$.

12. original claim: $\mu \geq 154$ [n ≤ 30 and σ unknown, use t]
   H_o: $\mu \geq 154$
   H_1: $\mu < 154$
   $\alpha = .005$
   C.R. t $< -t_{19,.005} = -2.861$
   calculations:
   $\quad t_{\bar{x}} = (\bar{x} - \mu)/s_{\bar{x}}$
   $\quad\quad = (141 - 154)/(12/\sqrt{20}\,)$
   $\quad\quad = -13/2.683$
   $\quad\quad = -4.845$
   conclusion:
   $\quad$ Reject H_o; there is sufficient evidence to conclude that $\mu < 154$.

13. The calculated t is larger in magnitude than the largest t (i.e., the t for the smallest α) in Table A-2 for its degrees of freedom. For this one-tailed test, therefore, conclude P-value $< .005$.

15. original claim: $\mu = 20.0$ [n ≤ 30 and σ unknown, use t]
   H_o: $\mu = 20.0$
   H_1: $\mu \neq 20.0$
   $\alpha = .02$
   C.R. t $< -t_{29,.01} = -2.462$
   $\quad\quad$ t $> t_{29,.01} = 2.462$
   calculations:
   $\quad t_{\bar{x}} = (\bar{x} - \mu)/s_{\bar{x}}$
   $\quad\quad = (20.5 - 20.0)/(1.5/\sqrt{30}\,)$
   $\quad\quad = .5/.274$
   $\quad\quad = 1.826$
   conclusion:
   $\quad$ Do not reject H_o; there is not sufficient evidence to conclude that $\mu \neq 20.0$.

17. original claim: $\mu > 2000$ [n > 30, use z (with s for σ)]
    H_o: $\mu \leq 2000$
    H_1: $\mu > 2000$
    $\alpha = .025$
    C.R. $z > z_{.025} = 1.960$
    calculations:

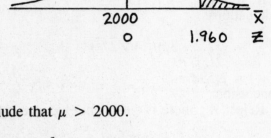

$$z_{\bar{x}} = (\bar{x} - \mu)/\sigma_{\bar{x}}$$
$$= (2177 - 2000)/(843/\sqrt{100})$$
$$= 177/84.3$$
$$= 2.100$$
    conclusion:
    Reject H_o; there is sufficient evidence to conclude that $\mu > 2000$.

18. original claim: $\mu < .88$ [n $\leq$ 30 and σ unknown, use t]
    H_o: $\mu \geq .88$
    H_1: $\mu < .88$
    $\alpha = .01$
    C.R. $t < -t_{7,.01} = -2.998$
    calculations:
$$t_{\bar{x}} = (\bar{x} - \mu)/s_{\bar{x}}$$
$$= (.76 - .88/(.04/\sqrt{8})$$
$$= -.12/.0141$$
$$= -8.458$$
    conclusion:
    Reject H_o; there is sufficient evidence to conclude that $\mu < .88$.

19. The calculated t is larger in magnitude than the largest t (i.e., the t for the smallest α) in Table A-2 for its degrees of freedom. For this one-tailed test, therefore, conclude P-value < .005.

21. original claim: $\mu = 243.5$ [n $\leq$ 30 and σ unknown, use t]
    H_o: $\mu = 243.5$
    H_1: $\mu \neq 243.5$
    $\alpha = .05$
    C.R. $t < -t_{21,.025} = -2.080$
    $\quad\quad t > t_{21,.025} = 2.080$
    calculations:
$$t_{\bar{x}} = (\bar{x} - \mu)/s_{\bar{x}}$$
$$= (170.2 - 243.5)/(35.3/\sqrt{22})$$
$$= -73.3/7.526$$
$$= -9.740$$
    conclusion:
    Reject H_o; there is sufficient evidence to conclude that $\mu \neq 243.5$ (in fact, $\mu < 243.5$).

23. original claim: $\mu > 5$ [n > 30, use z (with s for σ)]
    H_o: $\mu \leq 5$
    H_1: $\mu > 5$
    $\alpha = .10$
    C.R. $z > z_{.10} = 1.282$
    calculations:
$$z_{\bar{x}} = (\bar{x} - \mu)/\sigma_{\bar{x}}$$
$$= (5.15 - 5)/(1.68/\sqrt{80})$$
$$= .15/.188$$
$$= .799$$
    conclusion:
    Do not reject H_o; there is not sufficient evidence to conclude that $\mu > 5$.

25. original claim: $\mu = 650$ [n $\le$ 30 and σ unknown, use t]

H_o: $\mu = 650$

H_1: $\mu \ne 650$

$\alpha = .05$

C.R. t $<$ $-t_{29,.025} = -2.045$

     t $>$ $t_{29,.025} = 2.045$

calculations:

    $t_{\bar{x}} = (\bar{x} - \mu)/s_{\bar{x}}$

        $= (665.41 - 650)/(7.26/\sqrt{30})$

        $= 15.41/1.325$

        $= 11.626$

conclusion:

    Reject H_o; there is sufficient evidence to conclude that $\mu \ne 650$ (in fact, $\mu > 650$).

27. original claim: $\mu > 75$ [n $\le$ 30 and σ unknown, use t]

    n = 20

    $\Sigma x = 1561$ $\bar{x} = 78.05$

    $\Sigma x_2 = 122431$ $s = 5.596$

H_o: $\mu \le 75$

H_1: $\mu > 75$

$\alpha = .05$

C.R. t $>$ $t_{19,.05} = 1.729$

calculations:

    $t_{\bar{x}} = (\bar{x} - \mu)/s_{\bar{x}}$

        $= (78.05 - 75)/(5.596)/\sqrt{20})$

        $= 3.05/1.251$

        $= 2.438$

conclusion:

    Reject H_o; there is sufficient evidence to conclude that $\mu > 75$.

28. original claim: $\mu = 11,000$ [n $\le$ 30 and σ unknown, use t]

    n = 7

    $\Sigma x = 74,209$ $\bar{x} = 10,601.286$

    $\Sigma x_2 = 793,035,705$ $s = 1026.718$

H_o: $\mu = 11,000$

H_1: $\mu \ne 11,000$

$\alpha = .05$

C.R. t $<$ $-t_{6,.025} = -2.447$

      t $>$ $t_{6,.025} = 2.447$

calculations:

    $t_{\bar{x}} = (\bar{x} - \mu)/s_{\bar{x}}$

        $= (10,601.286 - 11,000)/(1026.718/\sqrt{7})$

        $= -398.714/388.063$

        $= -1.027$

conclusion:

    Do not reject H_o; there is not sufficient evidence to conclude that $\mu \ne 11,000$.

29. Since the calculated t is between $t_{6,.10} = 1.440$ and $t_{6,.25} = .718$, the P-value for this two-tailed test is between $2 \cdot (.10)$ and $2 \cdot (.25)$ -- i.e., $.20 <$ P-value $< .50$.

31. original claim: $\mu = 176$ [n ≤ 30 and σ unknown, use t]
    The following are the relevant entries from Data Set 3.

#	age	sex	married	weight
17	35	1	2	157
20	41	1	2	176
33	38	1	2	155
34	40	1	2	235
69	38	1	2	176
74	44	1	2	200
82	44	1	2	175
91	42	1	2	216
92	44	1	2	185
95	37	1	2	137
98	39	1	2	176
106	41	1	2	162
110	40	1	2	220
123	42	1	2	191
124	40	1	2	261

$n = 15$
$\Sigma x = 2822$
$\Sigma x^2 = 546,348$

$\bar{x} = 188.133$
$s = 33.205$

H_o: $\mu = 176$
H_1: $\mu \neq 176$
$\alpha = .05$
C.R. $t < -t_{14,.025} = -2.145$
     $t > t_{14,.025} = 2.145$
calculations:

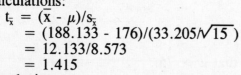

$t_{\bar{x}} = (\bar{x} - \mu)/s_{\bar{x}}$

    $= (188.133 - 176)/(33.205/\sqrt{15})$
    $= 12.133/8.573$
    $= 1.415$
conclusion:
    Do not reject H_o; there is not sufficient evidence to conclude that $\mu \neq 176$.

33. $A = z_{.05} \cdot [(8 \cdot df + 3)/(8 \cdot df + 1)]$
    $= 1.645 \cdot [(8 \cdot 9 + 3)/(8 \cdot 9 + 1)]$
    $= 1.645 \cdot [75/73]$
    $= 1.690$

$t = \sqrt{df \cdot (e^{A \cdot A/df} - 1)}$

    $= \sqrt{9 \cdot (e^{(1.690)(1.690)/9} - 1)}$

    $= \sqrt{9 \cdot (e^{.317} - 1)}$

    $= \sqrt{9 \cdot (.3735)}$

    $= 1.833$

This agrees exactly with $t_{9,.05} = 1.833$ given in Table A-3.

7-5 Testing a Claim about a Proportion

NOTE: To be consistent with the notation of the previous sections, and thereby reinforcing the patterns and concepts presented in those sections, the manual uses the "usual" z formula written to apply to $\hat{p}$'s

$$z_{\hat{p}} = (\hat{p} - \mu_{\hat{p}})/\sigma_{\hat{p}}$$

When the normal approximation to the binomial applies, the $\hat{p}$'s are normally distributed
with $\mu_{\hat{p}} = p$
and $\sigma_{\hat{p}} = \sqrt{pq/n}$

And so the formula for the z statistic may also be written as

$$z_{\hat{p}} = (\hat{p} - p)/\sqrt{pq/n}$$

1. original claim: $p \leq .04$ [normal approximation to the binomial, use z]
   $\hat{p} = x/n = 9/150 = .06$
   $H_0: p \leq .04$
   $H_1: p > .04$
   $\alpha = .05$
   C.R. $z > z_{.05} = 1.645$
   calculations:

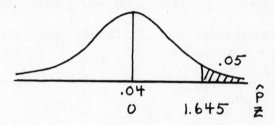

   $z_{\hat{p}} = (\hat{p} - \mu_{\hat{p}})/\sigma_{\hat{p}}$
   $= (.06 - .04)/\sqrt{(.04)(.96)/150}$
   $= .02/.016$
   $= 1.25$
   conclusion:
   Do not reject H_0; there is not sufficient evidence to conclude that $p > .04$.
   P-value $= P(z > 1.25) = .5000 - .3944 = .1056$
   We cannot be 95% certain that corrective action is needed.

3. original claim: $p = .71$ [normal approximation to the binomial, use z]
   $\hat{p} = x/n = .74$
   $H_0: p = .71$
   $H_1: p \neq .71$
   $\alpha = .10$
   C.R. $z < -z_{.05} = -1.645$
   $\quad\quad z > z_{.05} = 1.645$
   calculations:

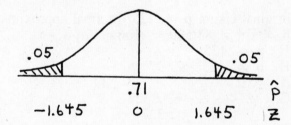

   $z_{\hat{p}} = (\hat{p} - \mu_{\hat{p}})/\sigma_{\hat{p}}$
   $= (.74 - .71)/\sqrt{(.71)(.29)/500}$
   $= .03/.0203$
   $= 1.478$
   conclusion:
   Do not reject H_0; there is not sufficient evidence to conclude that $p \neq .71$.
   P-value $= 2 \cdot P(z > 1.48) = 2 \cdot [.5000 - .4306] = 2 \cdot [.0694] = .1388$

5. original claim: $p = .20$ [normal approximation to the binomial, use z]
   $\hat{p} = x/n = 288/(288 + 962) = 288/1250 = .2304$
   H_o: $p = .20$
   H_1: $p \neq .20$
   $\alpha = .02$
   C.R. $z < -z_{.01} = -2.327$
   $z > z_{.01} = 2.327$
   calculations:

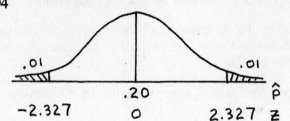

   $z_{\hat{p}} = (\hat{p} - \mu_{\hat{p}})/\sigma_{\hat{p}}$
   $= (.2304 - .20)/\sqrt{(.20)(.80)/1250}$
   $= .0304/.0113$
   $= 2.687$
   conclusion:
   Reject H_o; there is sufficient evidence to conclude that $p \neq .20$ (in fact, $p > .20$).
   P-value $= 2 \cdot P(z > 2.69) = 2 \cdot [.5000 - .4964] = 2 \cdot [.0036] = .0072$

7. original claim: $p > .25$ [normal approximation to the binomial, use z]
   $\hat{p} = x/n = .33$
   H_o: $p \leq .25$
   H_1: $p > .25$
   $\alpha = .02$
   C.R. $z > z_{.02} = 2.05$
   calculations:

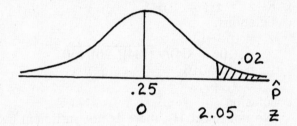

   $z_{\hat{p}} = (\hat{p} - \mu_{\hat{p}})/\sigma_{\hat{p}}$
   $= (.33 - .25)/\sqrt{(.25)(.75)/1400}$
   $= .08/.0116$
   $= 6.913$
   conclusion:
   Reject H_o; there is sufficient evidence to conclude that $p > .25$.
   P-value $= P(z > 6.91) = .5000 - .4999 = .0001$

9. original claim: $p = .125$ [normal approximation to the binomial, use z]
   $\hat{p} = x/n = 83/500 = .166$
   H_o: $p = .125$
   H_1: $p \neq .125$
   $\alpha = .02$
   C.R. $z < -z_{.01} = -2.327$
   $z > z_{.01} = 2.327$
   calculations:

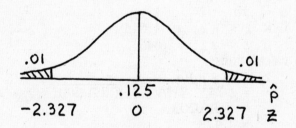

   $z_{\hat{p}} = (\hat{p} - \mu_{\hat{p}})/\sigma_{\hat{p}}$
   $= (.166 - .125)/\sqrt{(.125)(.875)/500}$
   $= .041/.0148$
   $= 2.772$
   conclusion:
   Reject H_o; there is sufficient evidence to conclude that $p \neq .125$ (in fact, $p > .125$).
   P-value $= 2 \cdot P(z > 12.77 = 2 \cdot [.5000 - .4972] = 2 \cdot [.0028] = .0056$

11. original claim: p > .08 [normal approximation to the binomial, use z]

$\hat{p} = x/n = 70/781 = .0896$

H_o: p ≤ .08

H_1: p > .08

$\alpha = .05$

C.R. z > $z_{.05}$ = 1.645

calculations

$z_{\hat{p}} = (\hat{p} - \mu_{\hat{p}})/\sigma_{\hat{p}}$

  $= (.0896 - .08)/\sqrt{(.08)(.92)/781}$

  $= .00962/.00971$

  $= .992$

conclusion:

   Do not reject H_o; there is not sufficient evidence to conclude that p > .08.

P-value = P(z > .99) = .5000 - .3389 = .1611

13. original claim: p < .10 [normal approximation to the binomial, use z]

$\hat{p} = x/n = 64/1068 = .0599$

H_o: p ≥ .10

H_1: p < .10

$\alpha = .01$

C.R. z < - $z_{.01}$ = -2.327

calculations

$z_{\hat{p}} = (\hat{p} - \mu_{\hat{p}})/\sigma_{\hat{p}}$

  $= (.0599 - .10)/\sqrt{(.10)(.90)/1068}$

  $= -.0401/.00918$

  $= -4.366$

conclusion:

   Reject H_o; there is sufficient evidence to conclude that p < .10.

P-value = P(z < -4.37) = .5000 - .43999 = .0001

15. original claim: p < .07 [normal approximation to the binomial, use z]

$\hat{p} = x/n = 333/5218 = .0638$

H_o: p ≥ .07

H_1: p < .07

$\alpha = .05$ [or any other appropriate value]

C.R. z < - $z_{.05}$ = -1.645

calculations

$z_{\hat{p}} = (\hat{p} - \mu_{\hat{p}})/\sigma_{\hat{p}}$

  $= (.0638 - .07)/\sqrt{(.07)(.93)/5218}$

  $= -.00618/.00353$

  $= -1.750$

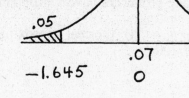

conclusion:

   Reject H_o; there is sufficient evidence to conclude that p < .07.

P-value = P(z < -1.75) = .5000 - .4599 = .0401

Statistically, we are 95% certain that the new system reduces the percent of no shows, but
   whether the estimated .6% reduction qualifies as "effective" must be decided by the airline
   executives.

17. original claim: p > .50 [normal approximation to the binomial, use z]
NOTE: A success is not being aware of the Holocaust.
$\hat{p} = x/n = 268/506 = .530$
$H_o: p \leq .50$
$H_1: p > .50$
$\alpha = .05$
C.R. $z > z_{.05} = 1.645$
calculations

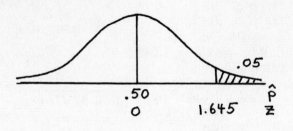

$z_{\hat{p}} = (\hat{p} - \mu_{\hat{p}})/\sigma_{\hat{p}}$
$= (.530 - .50)/\sqrt{(.50)(.50)/506}$
$= .0296/.0222$
$= 1.334$
conclusion:
Do not reject H_o; there is not sufficient evidence to conclude that p > .50.
P-value = P(z > 1.33) = .5000 - .4082 = .0918
We cannot be 95% certain that p is greater than 50% and that the curriculum needs to be revised.

19. original claim: p ≥ .50 [normal approximation to the binomial, use z]
$\hat{p} = x/n = .28$
$H_o: p \geq .50$
$H_1: p < .50$
$\alpha = .005$
C.R. $z < -z_{.005} = -2.575$
calculations

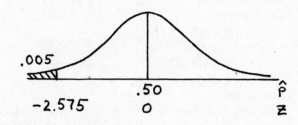

$z_{\hat{p}} = (\hat{p} - \mu_{\hat{p}})/\sigma_{\hat{p}}$
$= (.28 - .50)/\sqrt{(.50)(.50)/1012}$
$= -.22/.0157$
$= -13.997$
conclusion:
Reject H_o; there is sufficient evidence to conclude that p < .50.
P-value = P(z < -14.00) = .5000 - .4999 = .0001
The sample results are extremely unlikely to have occurred by chance if p ≥ .50. It appears that the fruitcake producers have a problem.

21. a. original claim: p = .20 [normal approximation to the binomial, use z]
$\hat{p} = x/n = 17/100 = .17$
$H_o: p = .20$
$H_1: p \neq .20$
$\alpha = .05$
C.R. $z < -z_{.025} = -1.960$
$z > z_{.025} = 1.960$
calculations

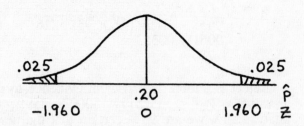

$z_{\hat{p}} = (\hat{p} - \mu_{\hat{p}})/\sigma_{\hat{p}}$
$= (.17 - .20)/\sqrt{(.20)(.80)/100}$
$= -.03/.04$
$= -.75$
conclusion:
Do not reject H_o; there is not sufficient evidence to conclude that p ≠ .20.

b. original claim: $p = .20$ [normal approximation to the binomial, use z]
$\hat{p} = x/n = 17/100 = .17$
$H_o: p = .20$
$H_1: p \neq .20$
$\alpha = .05$
C.R. P-value $< .05$
calculations

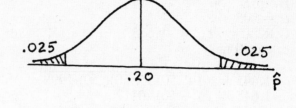

.025   .025

.20 $\hat{p}$

$z_{\hat{p}} = (\hat{p} - \mu_{\hat{p}})/\sigma_{\hat{p}}$
$= (.17 - .20)/\sqrt{(.20)(.80)/100}$
$= -.03/.04$
$= -.75$
P-value $= 2 \cdot P(z < -.75) = 2 \cdot [.5000 - .2734] = 2 \cdot [.2266] = .4532$
conclusion:
   Do not reject H_o; there is not sufficient evidence to conclude that $p \neq .20$.

c. There are 17 of the 100 M&M's that were red.
$\hat{p} = x/n$
$= 17/100$
$= .170$
$H_o: p = .20$
$H_1: p \neq .20$
$\alpha = .05$
calculations:
   $\hat{p} \pm z_{.025}\sqrt{\hat{p}\hat{q}/n}$
   $.170 \pm 1.960\sqrt{(.170)(.830)/100}$
   $.170 \pm .074$
   $.096 < p < .244$
conclusion:
   Since .20 is within the confidence interval, do not reject H_o; there is not sufficient
   evidence to conclude that $p \neq .20$.

23. original claim: $p = .10$ [$np = (15)(.10) = 1.5 < 5$; use binomial distribution]
   $x = 0$
                              NOTE: Refer to this probability
   $H_o: p = .10$ distribution for n=15 and p=.10.
   $H_1: p \neq .10$
   $\alpha = .05$
   C.R. $x < 0$ NOTE: obtained by placing .025 (or
        $x > 4$ much as possible) in each tail
   calculations
     $x = 0$
   conclusion:
     Do not reject H_o; there is not sufficient evidence
     to conclude that $p \neq .10$.
   P-value $= P($a value as extreme as our value$)$
          $= P(x=0 \text{ or } x \geq 3)$
          $= .206 + (.129 + .043 + .010 + .002 + 0^+ +...)$
          $= .206 + .184$
          $= .390$

x	P(x)
0	.206
1	.343
2	.267
3	.129
4	.043
5	.010
6	.002
7	0^+
.	.
.	.
.	.
	1.000

25. original claim: p < .50 [normal approximation to the binomial, use z]

H_o: p ≥ .50
H_1: p < .50
α = .05
C.R. z < $-z_{.05}$ = -1.645
calculations:

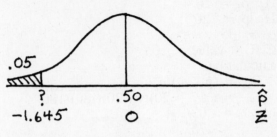

$z_{\hat{p}}$ = ($\hat{p}$ - $\mu_{\hat{p}}$)/$\sigma_{\hat{p}}$
-1.645 = ($\hat{p}$ - .50)/$\sqrt{(.50)(.50)/1998}$
-1.645 = ($\hat{p}$ - .50)/.0112
-.0184 = $\hat{p}$ - .50
.4816 = $\hat{p}$

NOTE: There is no conclusion because this exercise is not asking for the completion of a test of hypothesis. In fact, the sample proportion is irrelevant to the question posed. The above calculations find the $\hat{p}$ that corresponds to the z score that marks the beginning of the critical region. In terms of $\hat{p}$, then, the critical region is $\hat{p}$ < .4816.

The exercise asks for P(not rejecting H_o) = P($\hat{p}$ > .4816) when p = .45.

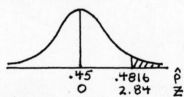

$z_{\hat{p}}$ = ($\hat{p}$ - $\mu_{\hat{p}}$)/$\sigma_{\hat{p}}$
= (.4816 - .45)/$\sqrt{(.45)(.55)/1998}$
= .0316/.0111
= 2.839
P(z > 2.84) = .5000 - .4977 = .0023

7-6 Testing a Claim about a Standard Deviation or Variance

1. a. $\chi^2_{19,.975}$ = 8.907
      $\chi^2_{19,.025}$ = 32.852

   b. $\chi^2_{19,.950}$ = 10.117

   c. $\chi^2_{74,.975}$ = 48.758 [closest entry]
         or 48.758 + (4/10)·(57.153 - 48.758) = 52.116 [using interpolation]

      $\chi^2_{74,.025}$ = 95.023 [closest entry]
         or 95.023 + (4/10)·(106.629 - 95.023) = 99.665 [using interpolation]

NOTE: Following the pattern used with the z and t distributions, this manual uses the closest entry from Table A-4 for χ^2 as if it were the precise value necessary and does not use interpolation. This procedure sacrifices very little accuracy -- and even interpolation does not yield precise values. When extreme accuracy is needed in practice, statisticians refer either to more accurate tables or to computer-produced values.

3. original claim: σ = 43.7
   H_o: σ = 43.7
   H_1: σ ≠ 43.7
   α = .05
   C.R. χ^2 < $\chi^2_{80,.975}$ = 57.153
        χ^2 > $\chi^2_{80,.025}$ = 106.629
   calculations:
      χ^2 = (n-1)s^2/σ^2
         = (80)(52.3)^2/(43.7)^2
         = 114.586
   conclusion:
      Reject H_o; there is sufficient evidence to conclude that σ ≠ 43.7 (in fact, σ > 43.7).

5. original claim: $\sigma < .75$
   H_o: $\sigma \geq .75$
   H_1: $\sigma < .75$
   $\alpha = .05$
   C.R. $\chi^2 < \chi^2_{60,.95} = 43.188$
   calculations:
   $\quad \chi^2 = (n-1)s^2/\sigma^2$
   $\qquad = (60)(.48)^2/(.75)^2$
   $\qquad = 24.576$
   conclusion:
   $\quad$ Reject H_o; there is sufficient evidence to conclude that $\sigma < 43.7$.

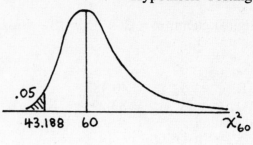

7. original claim: $\sigma < .15$
   H_o: $\sigma \geq .15$
   H_1: $\sigma < .15$
   $\alpha = .05$
   C.R. $\chi^2 < \chi^2_{70,.95} = 51.739$
   calculations:
   $\quad \chi^2 = (n-1)s^2/\sigma^2$
   $\qquad = (70)(.12)^2/(.15)^2$
   $\qquad = 44.800$
   conclusion:
   $\quad$ Reject H_o; there is sufficient evidence to conclude that $\sigma < .15$.
   $\quad$ Yes, since Medassist can be 95% certain that the new machine fills bottles with less variation they should consider its purchase.

9. original claim: $\sigma > 19.7$
   H_o: $\sigma \leq 19.7$
   H_1: $\sigma > 19.7$
   $\alpha = .05$
   C.R. $\chi^2 > \chi^2_{49,.05} = 67.505$
   calculations:
   $\quad \chi^2 = (n-1)s^2/\sigma^2$
   $\qquad = (49)(23.4)^2/(19.7)^2$
   $\qquad = 69.135$
   conclusion:
   $\quad$ Reject H_o; there is sufficient evidence to conclude that $\sigma > 19.7$.

11. original claim: $\sigma = 2.4$
    H_o: $\sigma = 2.4$
    H_1: $\sigma \neq 2.4$
    $\alpha = .05$
    C.R. $\chi^2 < \chi^2_{49,.975} = 32.357$
    $\qquad \chi^2 > \chi^2_{49,.025} = 71.420$
    calculations:
    $\quad \chi^2 = (n-1)s^2/\sigma^2$
    $\qquad = (49)(2.7)^2/(2.4)^2$
    $\qquad = 62.016$
    conclusion:
    $\quad$ Do not reject H_o; there is not sufficient evidence to conclude that $\sigma \neq 2.4$.

13. original claim: $\sigma = 3$
    H_o: $\sigma = 3$
    H_1: $\sigma \neq 3$
    $\alpha = .05$
    C.R. $\chi^2 < \chi^2_{61,.975} = 40.482$
           $\chi^2 > \chi^2_{61,.025} = 83.298$
    calculations:
        $\chi^2 = (n-1)s^2/\sigma^2$
           $= (61)(3.297)^2/(3)^2$
           $= 73.676$
conclusion:
    Do not reject H_o; there is not sufficient evidence to conclude that $\sigma \neq 3$.

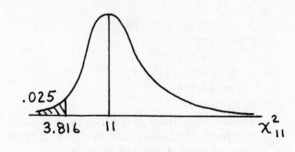

15. original claim: $\sigma < 2.0$
    $n = 12$
    $\Sigma x = 396.6$ $\bar{x} = 33.05$
    $\Sigma x^2 = 13121.66$ $s^2 = 1.275$
  H_o: $\sigma \geq 2.0$
  H_1: $\sigma < 2.0$
  $\alpha = .025$
  C.R. $\chi^2 < \chi^2_{11,.975} = 3.816$
  calculations:
      $\chi^2 = (n-1)s^2/\sigma^2$
         $= (11)(1.275)/(2.0)^2$
         $= 3.5075$
conclusion:
    Reject H_o; there is sufficient evidence to conclude that $\sigma < 2.0$.

17. original claim: $\sigma = .470$
    $n = 16$
    $\Sigma x = 58.8$ $\bar{x} = 3.675$
    $\Sigma x^2 = 222.571$ $s^2 = .432$
  H_o: $\sigma = .470$
  H_1: $\sigma \neq .470$
  $\alpha = .05$ [or any other appropriate value]
  C.R. $\chi^2 < \chi^2_{15,.975} = 6.262$
        $\chi^2 > \chi^2_{15,.025} = 27.488$
  calculations:
      $\chi^2 = (n-1)s^2/\sigma^2$
         $= (15)(.432)/(.470)^2$
         $= 29.339$
conclusion:
    Reject H_o; there is sufficient evidence to conclude that $\sigma \neq .470$ (in fact, $\sigma > .470$).

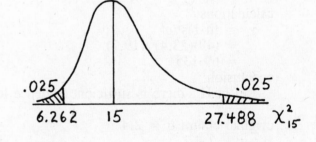

19. a. To find the P-value for $\chi^2_9 = 19.735$ in a one-tailed greater than test,
    note that $\chi^2_{9,.025} = 19.023$ and $\chi^2_{9,.01} = 21.666$.
    Therefore, $.01 < \text{P-value} < .025$.

  b. To find the P-value for $\chi^2_{19} = 7.337$ in a one-tailed less than test,
    note that $\chi^2_{19,.995} = 6.844$ and $\chi^2_{19,.99} = 7.633$.
    Therefore, $.005 < \text{P-value} < .01$.

c. To find the P-value for $\chi^2_{29} = 54.603$ in a two-tailed test,
note that $\chi^2_{29,.005} = 52.336$ is the largest χ^2 value in the row.
Therefore, P-value $< 2 \cdot (.005)$
P-value $< .01$

21. a. upper $\chi^2 = \frac{1}{2}(z_{.025} + \sqrt{2 \cdot df - 1})^2$
$= \frac{1}{2}(1.960 + \sqrt{2 \cdot (100) - 1})^2$
$= \frac{1}{2}(1.960 + \sqrt{199})^2$
$= \frac{1}{2}(258.140)$
$= 129.070$ [compare to $\chi^2_{100,.025} = 129.561$ from Table A-4]
lower $\chi^2 = \frac{1}{2}(-z_{.025} + \sqrt{2 \cdot df - 1})^2$
$= \frac{1}{2}(-1.960 + \sqrt{2 \cdot (100) - 1})^2$
$= \frac{1}{2}(-1.960 + \sqrt{199})^2$
$= \frac{1}{2}(147.543)$
$= 73.772$ [compare to $\chi^2_{100,.975} = 74.222$ from Table A-4]

b. upper $\chi^2 = \frac{1}{2}(z_{.025} + \sqrt{2 \cdot df - 1})^2$
$= \frac{1}{2}(1.960 + \sqrt{2 \cdot (149) - 1})^2$
$= \frac{1}{2}(1.960 + \sqrt{297})^2$
$= \frac{1}{2}(368.398)$
$= 184.199$
lower $\chi^2 = \frac{1}{2}(-z_{.025} + \sqrt{2 \cdot df - 1})^2$
$= \frac{1}{2}(-1.960 + \sqrt{2 \cdot (149) - 1})^2$
$= \frac{1}{2}(-1.960 + \sqrt{297})^2$
$= \frac{1}{2}(233.286)$
$= 116.643$

23. original claim: $\sigma < 6.2$
H_o: $\sigma \geq 6.2$
H_1: $\sigma < 6.2$
$\alpha = .05$
C.R. $\chi^2 < \chi^2_{24,.95} = 13.848$
calculations:
$\chi^2 = (n-1)s^2/\sigma^2$
$13.848 = (24)s^2/(6.2)^2$
$532.317 = (24)s^2$
$22.180 = s^2$

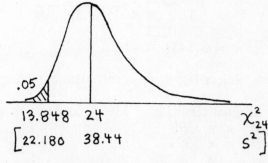

NOTE: There is no conclusion because this exercise is not asking for the completion of a test of hypothesis. In fact, the sample data is irrelevant to the question posed. The above calculations find the s^2 that corresponds to the χ^2 score that marks the beginning of the critical region. In terms of s^2, then, the critical region is $s^2 < 22.180$

The exercise asks for P(not rejecting H_o) = P($s^2 > 22.180$) when $\sigma = 4.0$.
$\chi^2 = (n-1)s^2/\sigma^2$
$= (24)(22.180)/(4.0)^2$
$= 33.270$
P($\chi^2_{24} > 33.270$) = .10   [from Table A-4]

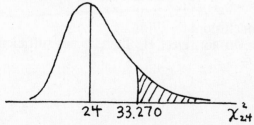

Review Exercises

1. a. concerns p: normal approximation to the binomial, use z

   $-z_{.05} = -1.645$

   b. concerns μ: n > 30, use z (with s for σ)

   $z_{.01} = 2.327$

   c. concerns μ: n < 30 and σ unknown, use t

   $\pm t_{11,.005} = \pm 3.106$

   d. concerns σ, use χ^2

   $\chi^2_{24,.99} = 10.856$

   e. concerns σ, use χ^2

   $\chi^2_{29,.975} = 16.047$ and $\chi^2_{29,.025} = 45.772$

3. a. H_o: $\mu \geq 20.0$

   b. left-tailed [since the alternative hypothesis is H_1: $\mu < 20.0$]

   c. concluding that the average treatment time is less than 20.0 when it really is not

   d. failing to conclude that the average treatment time is less than 20.0 when it really is

   e. $\alpha = .01$

5. original claim: p > .50 [concerns p: normal approximation to the binomial, use z]

   $\hat{p} = x/n = .57$

   H_o: p ≤ .50

   H_1: p > .50

   $\alpha = .01$

   C.R. z > $z_{.01} = 2.327$

   calculations:

   $z_{\hat{p}} = (\hat{p} - \mu_{\hat{p}})/\sigma_{\hat{p}}$
   $= (.57 - .50)/\sqrt{(.50)(.50)/504}$
   $= .07/.0223$
   $= 3.143$

   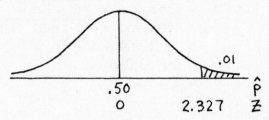

   conclusion:

   Reject H_o; there is sufficient evidence to conclude that p > .50.

7. original claim: $\mu = 3393$ [concerns μ: n ≤ 30 and σ unknown, use t]

   H_o: $\mu = 3393$

   H_1: $\mu \neq 3393$

   $\alpha = .05$

   C.R. t < $-t_{29,.025} = -2.045$

   t > $t_{29,.025} = 2.045$

   calculations:

   $t_{\bar{x}} = (\bar{x} - \mu)/s_{\bar{x}}$
   $= (3264 - 3393)/(485/\sqrt{30})$
   $= -129/88.548$
   $= -1.457$

   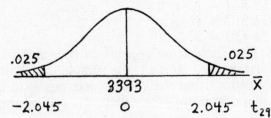

   conclusion:

   Do not reject H_o; there is not sufficient evidence to conclude that $\mu \neq 3393$.

9. original claim: $\mu < 5.00$ [concerns μ: n > 30, use z (with s for σ)]
  H_o: $\mu \geq 5.00$
  H_1: $\mu < 5.00$
  $\alpha = .01$
  C.R. z < $-z_{.01}$ = -2.327
  calculations:

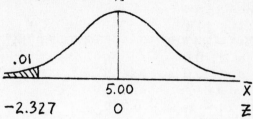

  $z_{\bar{x}} = (\bar{x} - \mu)/\sigma_{\bar{x}}$
  $= (4.13 - 5.00)/(1.91/\sqrt{36}\,)$
  $= -.87/.3183$
  $= -2.733$
  conclusion:
    Reject H_o; there is sufficient evidence to conclude that $\mu < 5.00$.

11. original claim: $\mu > 55.0$ [concerns μ: n > 30, use z (with s for σ)]
  H_o: $\mu \leq 55.0$
  H_1: $\mu > 55.0$
  $\alpha = .025$
  C.R. z > $z_{.025}$ = 1.960
  calculations:

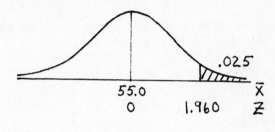

  $z_{\bar{x}} = (\bar{x} - \mu)/\sigma_{\bar{x}}$
  $= (61.3 - 55.0)/(3.3/\sqrt{50}\,)$
  $= 6.3/.4667$
  $= 13.499$
  conclusion:
    Reject H_o; there is sufficient evidence to conclude that $\mu > 55.0$.

13. original claim: $\sigma^2 > 6410$ [concerns σ: use χ^2]
  H_o: $\sigma^2 \leq 6410$
  H_1: $\sigma^2 > 6410$
  $\alpha = .10$
  C.R. $\chi^2 > \chi^2_{59,.10}$ = 74.397
  calculations:

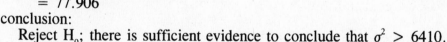

  $\chi^2 = (n-1)s^2/\sigma^2$
  $= (59)(8464)/(6410)$
  $= 77.906$
  conclusion:
    Reject H_o; there is sufficient evidence to conclude that $\sigma^2 > 6410$.

15. original claim: p = .98 [concerns p: normal approximation to the binomial, use z]
  $\hat{p} = x/n = 831/849 = .9788$
  H_o: p = .98
  H_1: p $\neq$ .98
  $\alpha = .10$
  C.R. z < $- z_{.05}$ = -1.645
    z > $z_{.05}$ = 1.645
  calculations:

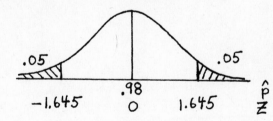

  $z_{\hat{p}} = (\hat{p} - \mu_{\hat{p}})/\sigma_{\hat{p}}$
  $= (.9788 - .98)/\sqrt{(.98)(.02)/849}$
  $= -.0012/.00480$
  $= -.250$
  conclusion:
    Do not reject H_o; there is not sufficient evidence to conclude that p $\neq$ .98.

17. original claim: $\sigma < 10.0$ [concerns σ: use χ^2]

    n = 15

    $\Sigma x = 1470$ $\bar{x} = 98.0$

    $\Sigma x^2 = 145050$ $s^2 = 70.714$

H_o: $\sigma \geq 10.0$

H_1: $\sigma < 10.0$

$\alpha = .05$

C.R. $\chi^2 < \chi^2_{14,.95} = 6.571$

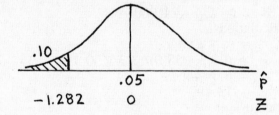

calculations:

    $\chi^2 = (n-1)s^2/\sigma^2$

       $= (14)(70.714)/(10)^2$

       $= 9.90$

conclusion:

    Do not reject H_o; there is not sufficient evidence to conclude that $\sigma < 10.0$.

19. original claim: $p < .05$ [concerns p: normal approximation to the binomial, use z]

    $\hat{p} = x/n = 72/1500 = .048$

H_o: $p \geq .05$

H_1: $p < .05$

$\alpha = .10$

C.R. $z < -z_{.10} = -1.282$

calculations:

    $z_{\hat{p}} = (\hat{p} - \mu_{\hat{p}})/\sigma_{\hat{p}}$

       $= (.048 - .05)/\sqrt{(.05)(.95)/1500}$

       $= -.002/.005627$

       $= -.355$

conclusion:

    Do not reject H_o; there is not sufficient evidence to conclude that $p < .05$.

P-value $= P(z < -.36) = .5000 - .1406 = .3594$

Chapter 8

Inferences from Two Samples

8-2 Inferences about Two Means: Dependent Samples

NOTE: To be consistent with the notation of the previous sections, and thereby reinforcing the patterns and concepts presented in those sections, the manual uses the "usual" t formula written to apply to $\bar{d}$'s

$$t_{\bar{d}} = (\bar{d} - \mu_{\bar{d}})/s_{\bar{d}}$$

with $\mu_{\bar{d}} = \mu_d$

and $s_{\bar{d}} = s_d/\sqrt{n}$

And so the formula for the t statistic may also be written as

$$t_{\bar{d}} = (\bar{d} - \mu_d)/(s_d/\sqrt{n})$$

1. d = x - y: 1 -1 0 1 -1 2 3 3 2 2 3
   $n = 11$
   $\Sigma d = 15$
   $\Sigma d^2 = 43$
   a. $\bar{d} = (\Sigma d)/n$
      $= 15/11$
      $= 1.363$
   b. $s_d^2 = [n \cdot \Sigma d^2 - (\Sigma d)^2]/[n(n-1)]$
      $= [11 \cdot 43 - (15)^2]/[11(10)]$
      $= 248/110$
      $= 2.225$
      $s_d = 1.502$
   c. $t_{\bar{d}} = (\bar{d} - \mu_{\bar{d}})/s_{\bar{d}}$
      $= (1.364 - 0)/(1.502/\sqrt{11})$
      $= 1.363/.4527$
      $= 3.012$
   d. $\pm t_{10,.025} = \pm 2.228$

3. $\bar{d} \pm t_{10,.025} \cdot s_d/\sqrt{n}$
   $1.4 \pm 2.228 \cdot 1.502/\sqrt{11}$
   1.4 ± 1.0
   $.4 < \mu_d < 2.4$

5. original claim $\mu_d = 0$ [$n \le 30$ and σ_d unknown, use t]
   $d = x_H - x_S$: 2.1 -1.0 1.7 1.6 1.4 2.5 1.3 2.6
     $n = 8$
      $\Sigma d = 12.2$ $\bar{d} = 1.525$
      $\Sigma d^2 = 27.52$ $s_d = 1.129$
   H_o: $\mu_d = 0$
   H_1: $\mu_d \ne 0$
   $\alpha = .05$
   C.R. $t < -t_{7,.025} = -2.365$
        $t > t_{7,.025} = 2.365$
   calculations:

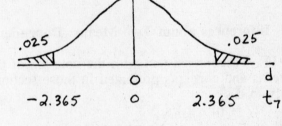

     $t_{\bar{d}} = (\bar{d} - \mu_{\bar{d}})/s_{\bar{d}}$
        $= (1.525 - 0)/(1.129/\sqrt{8})$
        $= 1.525/.399$
        $= 3.822$
   conclusion:
     Reject H_o; there is sufficient evidence to conclude that $\mu_d \ne 0$ (in fact, $\mu_d > 0$).

7. $d = x_B - x_A$: 2 14 17 -7 17 12 0 15 33 16
     $n = 10$
      $\Sigma d = 119$ $\bar{d} = 11.9$
      $\Sigma d^2 = 2541$ $s_d = 11.180$
   $\bar{d} \pm t_{9,.025} \cdot s_d/\sqrt{n}$
   $11.9 \pm 2.262 \cdot 11.180/\sqrt{10}$
   11.9 ± 8.0
   $3.9 < \mu_d < 19.9$

9. original claim $\mu_d > 0$ [$n \le 30$ and σ_d unknown, use t]
   $d = x_B - x_A$: -.2 4.1 1.6 1.8 3.2 2.0 2.9 9.6
     $n = 8$
      $\Sigma d = 25.0$ $\bar{d} = 3.125$
      $\Sigma d^2 = 137.46$ $s_d = 2.911$
   H_o: $\mu_d \le 0$
   H_1: $\mu_d > 0$
   $\alpha = .05$
   C.R. $t > t_{7,.05} = 1.895$
   calculations:
     $t_{\bar{d}} = (\bar{d} - \mu_{\bar{d}})/s_{\bar{d}}$
        $= (3.125 - 0)/(2.911/\sqrt{8})$
        $= 3.125/1.029$
        $= 3.036$
   conclusion:
     Reject H_o; there is sufficient evidence to conclude that $\mu_d > 0$.

11. $d = x_B - x_A$: -36 10 15 -2 0 -22 -36 -6 -26 7
     $n = 10$
      $\Sigma d = -96$ $\bar{d} = -9.6$
      $\Sigma d^2 = 4106$ $s_d = 18.987$
   $\bar{d} \pm t_{9,.025} \cdot s_d/\sqrt{n}$
   $-9.6 \pm 2.262 \cdot 18.987/\sqrt{10}$
   -9.6 ± 13.6
   $-23.2 < \mu_d < 4.0$

13. original claim $\mu_d = 0$ [n ≤ 30 and σ_d unknown, use t]
  d = x_{pre} - x_{post}: 5 0 0 0 8 1 1 4 0 1
    n = 10
    Σd = 20 $\bar{d}$ = 2.0
    Σd^2 = 108 s_d = 2.749
  H_o: $\mu_d = 0$
  H_1: $\mu_d \neq 0$
  α = .05
  C.R. t < -$t_{9,.025}$ = -2.262
       t > $t_{9,.025}$ = 2.262
  calculations:

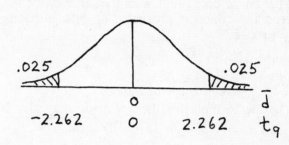

$$t_{\bar{d}} = (\bar{d} - \mu_{\bar{d}})/s_{\bar{d}}$$
$$= (2.0 - 0)/(2.749/\sqrt{10})$$
$$= 2.0/.869$$
$$= 2.301$$
  conclusion:
    Reject H_o; there is sufficient evidence to conclude that $\mu_d \neq 0$ (in fact, $\mu_d > 0$).
  We can be 95% certain that physical training does affect weight (and that the weight is less after the training).

15. original claim $\mu_d = 0$ [n ≤ 30 and σ_d unknown, use t]
  d = x_8 - x_{12}: -.7 -.8 -1.6 -.5 .8 .1 -.5 0 -1.3 -1.8 -2.3
    n = 11
    Σd = -8.6 $\bar{d}$ = -.782
    Σd^2 = 15.06 s_d = .913
  H_o: $\mu_d = 0$
  H_1: $\mu_d \neq 0$
  α = .05
  C.R. t < -$t_{10,.025}$ = -2.228
       t > $t_{10,.025}$ = 2.228
  calculations:
$$t_{\bar{d}} = (\bar{d} - \mu_{\bar{d}})/s_{\bar{d}}$$
$$= (-.782 - 0)/(.913/\sqrt{11})$$
$$= -.782/.275$$
$$= -2.840$$
  conclusion:
    Reject H_o; there is sufficient evidence to conclude that $\mu_d \neq 0$ (in fact, $\mu_d < 0$).
  No, it appears that a female's noon temperature is higher than her morning temperature.

17. a. To find the P-value for $t_7 = 3.822$ in a two-tailed test,
    note that $t_{7,.005} = 3.500$ is the largest t value in the row.
    Therefore, P-value < 2·(.005)
             P-value < .01

  b. To find the P-value for $t_7 = 3.036$ in a one-tailed greater than test,
    note that $t_{7,.01} = 2.998$ and $t_{7,.005} = 3.500$.
    Therefore, .005 < P-value < .01.

19. The $0 < \mu_d < 1.2$ confidence interval was obtained from $\bar{d} \pm t_{df,.025} \cdot s_d/\sqrt{n}$,
  which corresponds to a test of hypothesis critical region
    C.R. t < -$t_{df,.025}$
       t > $t_{df,.025}$.
  The one-tailed greater than test of hypothesis with the same relevant critical boundary has
    C.R. t > $t_{df,.025}$
  which corresponds to α = .025.

8-3 Inferences about Two Means: Independent and Large Samples

NOTE: To be consistent with the notation of the previous sections, and thereby reinforcing the patterns and concepts presented in those sections, the manual uses the "usual" z formula written to apply to $\bar{x}_1$-$\bar{x}_2$'s

$$z_{\bar{x}_1-\bar{x}_2} = (\bar{x}_1-\bar{x}_2 - \mu_{\bar{x}_1-\bar{x}_2})/\sigma_{\bar{x}_1-\bar{x}_2}$$

with $\mu_{\bar{x}_1-\bar{x}_2} = \mu_1 - \mu_2$

 and $\sigma_{\bar{x}_1-\bar{x}_2} = \sqrt{\sigma_1^2/n_1 + \sigma_2^2/n_2}$

And so the formula for the z statistic may also be written as

$$z_{\bar{x}_1-\bar{x}_2} = ((\bar{x}_1-\bar{x}_2) - (\mu_1-\mu_2))/\sqrt{\sigma_1^2/n_1 + \sigma_2^2/n_2}$$

1. original claim: $\mu_1-\mu_2 = 0$ [$n_1 > 30$ and $n_2 > 30$, use z (with s's for σ's)]
    $\bar{x}_1-\bar{x}_2 = 79.6 - 84.2 = -4.6$
    H_o: $\mu_1-\mu_2 = 0$
    H_1: $\mu_1-\mu_2 \neq 0$
    $\alpha = .05$
    C.R. $z < -z_{.025} = -1.960$
    $\quad z > z_{.025} = 1.960$
    calculations:

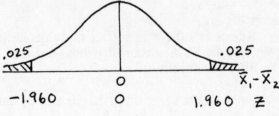

$$z_{\bar{x}_1-\bar{x}_2} = (\bar{x}_1-\bar{x}_2 - \mu_{\bar{x}_1-\bar{x}_2})/\sigma_{\bar{x}_1-\bar{x}_2}$$
$$= (-4.6 - 0)/\sqrt{(12.4)^2/40 + (12.2)^2/40}$$
$$= -4.6/2.750$$
$$= -1.672$$

conclusion:
    Do not reject H_o; there is not sufficient evidence to conclude that $\mu_1-\mu_2 \neq 0$.

3. $(\bar{x}_1-\bar{x}_2) \pm z_{.025}\sqrt{\sigma_1^2/n_1 + \sigma_2^2/n_2}$
    $-4.6 \pm 1.960 \cdot \sqrt{(12.4)^2/40 + (12.2)^2/40}$
    -4.6 ± 5.4
    $-10.0 < \mu_1-\mu_2 < .8$

5. Let the males be group 1.
    original claim: $\mu_1-\mu_2 > 0$ [$n_1 > 30$ and $n_2 > 30$, use z (with s's for σ's)]
    $\bar{x}_1-\bar{x}_2 = 35,330 - 29,610 = 5720$
    H_o: $\mu_1-\mu_2 \leq 0$
    H_1: $\mu_1-\mu_2 > 0$
    $\alpha = .01$
    C.R. $z > z_{.01} = 2.327$
    calculations:

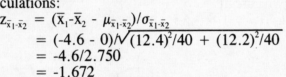

$$z_{\bar{x}_1-\bar{x}_2} = (\bar{x}_1-\bar{x}_2 - \mu_{\bar{x}_1-\bar{x}_2})/\sigma_{\bar{x}_1-\bar{x}_2}$$
$$= (5720 - 0)/\sqrt{(23,260)^2/86 + (16,480)^2/39}$$
$$= 5720/3640.72$$
$$= 1.571$$

conclusion:
    Do not reject H_o; there is not sufficient evidence to conclude that $\mu_1-\mu_2 > 0$.

7. Let the men aged 25-34 be group 1.

$(\bar{x}_1 - \bar{x}_2) \pm z_{.005}\sqrt{\sigma_1^2/n_1 + \sigma_2^2/n_2}$

$(176 - 164) \pm 2.575 \cdot \sqrt{(35.0)^2/804 + (27.0)^2/1657}$

12 ± 4

$8 < \mu_1 - \mu_2 < 16$

NOTE: Since $\bar{x}_1 = 176$ and $\bar{x}_2 = 164$ are given only to the nearest pound, and the value of any decimal places is unknown, more accuracy than that cannot be given in the answer.

9. Let American be group 1.

original claim: $\mu_1 - \mu_2 = 0$ [$n_1 > 30$ and $n_2 > 30$, use z (with s's for σ's)]

$\bar{x}_1 - \bar{x}_2 = 23{,}870 - 22{,}025 = 1845$

H_o: $\mu_1 - \mu_2 = 0$

H_1: $\mu_1 - \mu_2 \neq 0$

$\alpha = .10$

C.R. $z < -z_{.05} = -1.645$

$\quad\quad z > z_{.05} = 1.645$

calculations:

$z_{\bar{x}_1 - \bar{x}_2} = (\bar{x}_1 - \bar{x}_2 - \mu_{\bar{x}_1 - \bar{x}_2})/\sigma_{\bar{x}_1 - \bar{x}_2}$

$\quad\quad = (1845 - 0)/\sqrt{(2960)^2/40 + (3065)^2/35}$

$\quad\quad = 1845/698.17$

$\quad\quad = 2.643$

conclusion:

Reject H_o; there is sufficient evidence to conclude that $\mu_1 - \mu_2 \neq 0$ (in fact, $\mu_1 - \mu_2 > 0$). The conclusion that the mean pay is more at American than TWA is not necessarily relevant for an individual considering employment. Both companies might have exactly the same pay scale, for example, but the American attendants might have more accumulated experience and hence be paid more.

11. Let Tom's class be group 1.

original claim: $\mu_1 - \mu_2 = 0$ [$n_1 > 30$ and $n_2 > 30$, use z (with s's for σ's)]

$\bar{x}_1 - \bar{x}_2 = 76.0 - 73.4 = 2.6$

H_o: $\mu_1 - \mu_2 = 0$

H_1: $\mu_1 - \mu_2 \neq 0$

$\alpha = .05$

C.R. $z < -z_{.025} = -1.960$

$\quad\quad z > z_{.025} = 1.960$

calculations:

$z_{\bar{x}_1 - \bar{x}_2} = (\bar{x}_1 - \bar{x}_2 - \mu_{\bar{x}_1 - \bar{x}_2})/\sigma_{\bar{x}_1 - \bar{x}_2}$

$\quad\quad = (2.6 - 0)/\sqrt{(14.1)^2/35 + (13.5)^2/40}$

$\quad\quad = 2.6/3.199$

$\quad\quad = .812$

conclusion:

Do not reject H_o; there is not sufficient evidence to conclude that $\mu_1 - \mu_2 \neq 0$.

13. Let the East be group 1.

$(\bar{x}_1 - \bar{x}_2) \pm z_{.025}\sqrt{\sigma_1^2/n_1 + \sigma_2^2/n_2}$

$(421 - 347) \pm 1.960 \cdot \sqrt{(122)^2/35 + (85)^2/50}$

74 ± 47

$27 < \mu_1 - \mu_2 < 121$

No, the confidence interval does not include zero. This suggests that there is a significant difference between the two means (in fact, that μ_{East} is larger than μ_{West}).

15. Let Weston be group 1.

original claim: $\mu_1-\mu_2 = 0$ [$n_1 > 30$ and $n_2 > 30$, use z (with s's for σ's)]

$\bar{x}_1-\bar{x}_2 = 12.2 - 14.0 = -1.8$

H_o: $\mu_1-\mu_2 = 0$

H_1: $\mu_1-\mu_2 \neq 0$

$\alpha = .05$

C.R. z $< -z_{.025} = -1.960$

  z $> z_{.025} = 1.960$

calculations:

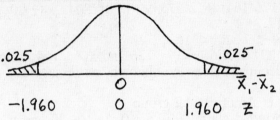

$$z_{\bar{x}_1-\bar{x}_2} = (\bar{x}_1-\bar{x}_2 - \mu_{\bar{x}_1-\bar{x}_2})/\sigma_{\bar{x}_1-\bar{x}_2}$$
$$= (-1.8 - 0)/\sqrt{(1.5)^2/50 + (2.1)^2/50}$$
$$= -1.8/.365$$
$$= -4.932$$

conclusion:

Reject H_o; there is sufficient evidence to conclude that $\mu_1-\mu_2 \neq 0$ (in fact, $\mu_1-\mu_2 < 0$). The conclusion that the mean response time is less for Weston than for Mid-Valley is not necessarily relevant for an individual considering calling an ambulance. Both companies might have exactly the same response time over the same distance, for example, but the Weston sample might have involved smaller distances. Similarly, the most important criterion for any one particular person in need of an ambulance to consider would probably be the distance from the ambulance company to the scene of the emergency.

17. Let the 18-24 year olds be group 1.

original claim: $\mu_1-\mu_2 = 0$ [$n_1 > 30$ and $n_2 > 30$, use z (with s's for σ's)]

group 1: 18-24 inclusive group 2: 25 and older

n = 37   n = 56

Σx = 3634.3 Σx = 5491.2

Σx^2 = 356,992.32 Σx^2 = 538,473.56

$\bar{x}$ = 98.2243 $\bar{x}$ = 98.0571

s^2 = .4349 s^2 = .4032

$\bar{x}_1-\bar{x}_2 = 98.2243 - 98.0571 = .1672$

a. H_o: $\mu_1-\mu_2 = 0$

   H_1: $\mu_1-\mu_2 \neq 0$

   $\alpha = .05$

   C.R. z $< -z_{.025} = -1.960$

     z $> z_{.025} = 1.960$

   calculations:

$$z_{\bar{x}_1-\bar{x}_2} = (\bar{X}_1-\bar{X}_2 - \mu_{\bar{x}_1-\bar{x}_2})/\sigma_{\bar{x}_1-\bar{x}_2}$$
$$= (.1672 - 0)/\sqrt{.4349/37 + .4032/56}$$
$$= .1672/.1377$$
$$= 1.214$$

conclusion:

Do not reject H_o; there is not sufficient evidence to conclude that $\mu_1-\mu_2 \neq 0$.

b. H_o: $\mu_1\text{-}\mu_2 = 0$
   H_1: $\mu_1\text{-}\mu_2 \neq 0$
   $\alpha = .05$
   C.R. P-value $< .05$
   calculations:

$$z_{\bar{x}_1\text{-}\bar{x}_2} = (\bar{x}_1\text{-}\bar{x}_2 - \mu_{\bar{x}_1\text{-}\bar{x}_2})/\sigma_{\bar{x}_1\text{-}\bar{x}_2}$$
$$= (.1672 - 0)/\sqrt{.4349/37 + .4032/56}$$
$$= .1672/.1377$$
$$= 1.214$$

   P-value $= 2 \cdot P(z > 1.21) = 2 \cdot (.5000 - .3869) = 2 \cdot (.1131) = .2262$
   conclusion:
   Do not reject H_o; there is not sufficient evidence to conclude that $\mu_1\text{-}\mu_2 \neq 0$.

c. $(\bar{x}_1\text{-}\bar{x}_2) \pm z_{.025}\sqrt{\sigma_1^2/n_1 + \sigma_2^2/n_2}$
   $(98.2243 - 98.0571) \pm 1.960 \cdot \sqrt{.4349/37 + .4032/56}$
   $.17 \pm .27$
   $-.10 < \mu_1\text{-}\mu_2 < .44$
   Since the confidence interval includes the value zero, there is not sufficient evidence to conclude that $\mu_1\text{-}\mu_2 \neq 0$.

19. a. $x = 5,10,15$
    $\mu = \Sigma x/n = 30/3 = 10$
    $\sigma^2 = \Sigma(x\text{-}\mu)^2/n$
    $= [(-5)^2 + (0)^2 + (5)^2]/3$
    $= 50/3$

b. $y = 1,2,3$
   $\mu = \Sigma y/n = 6/3 = 2$
   $\sigma^2 = \Sigma(y\text{-}\mu)^2/n$
   $= [(-1)^2 + (0)^2 + (1)^2]/3$
   $= 2/3$

c. $z = x\text{-}y = 4,3,2,9,8,7,14,13,12$
   $\mu = \Sigma z/n = 72/9 = 8$
   $\sigma^2 = \Sigma(z\text{-}\mu)^2/n$
   $= [(-4)^2 + (-5)^2 + (-6)^2 + (1)^2 + (0)^2 + (-1)^2 + (6)^2 + (5)^2 + (4)^2]/9$
   $= 156/9$
   $= 52/3$

d. $\sigma_{x\text{-}y}^2 = \sigma_x^2 + \sigma_y^2$
   $52/3 = 50/3 + 2/3$
   $52/3 = 52/3$

8-4 Comparing Two Variances

NOTE: The following conventions are used in this manual regarding the F test.
* The set of scores with the larger sample variance is designated group 1.
* Even though always designating the scores with the larger sample variance as group 1 makes lower critical values are unnecessary in two-tailed tests, the lower critical value will be calculated (using the method given in exercise #19) and included (in brackets) for completeness and for consistency with the other tests.
* The degrees of freedom for group 1 (numerator) and group 2 (denominator) will be given with the F as a superscript and subscript respectively.
* If the desired degrees of freedom does not appear in Table A-5, the closest entry will be used. If the desired degrees of freedom is exactly halfway between two tabled values, the conservative approach of using the smaller degrees of freedom is employed. Since any finite number is closer to 120 than ∞, 120 is used for all degrees of freedom larger than 120.
* Since all hypotheses in the text question the equality of σ_1^2 and σ_2^2, the calculation of F [which is statistically defined to be $F = (s_1^2/\sigma_1^2)/(s_2^2/\sigma_2^2)$] is shortened to $F = s_1^2/s_2^2$.

1. Let sample A be group 1.
   original claim: $\sigma_1^2 = \sigma_2^2$
   H_o: $\sigma_1^2 = \sigma_2^2$
   H_1: $\sigma_1^2 \neq \sigma_2^2$
   $\alpha = .05$
   C.R. $F < F_{9,.975}^{9} = [.2484]$
   $\qquad F > F_{9,.025}^{9} = 4.0260$
   calculations:
   $\quad F = s_1^2/s_2^2$
   $\qquad = 50/25$
   $\qquad = 2.0000$
   conclusion:
   Do not reject H_o; there is not sufficient evidence to conclude that $\sigma_1^2 \neq \sigma_2^2$.

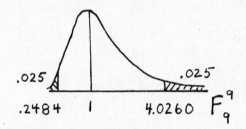

3. Let sample A be group 1.
   original claim: $\sigma_1^2 > \sigma_2^2$
   H_o: $\sigma_1^2 \leq \sigma_2^2$
   H_1: $\sigma_1^2 > \sigma_2^2$
   $\alpha = .05$
   C.R. $F > F_{19,.05}^{15} = 1.7505$
   calculations:
   $\quad F = s_1^2/s_2^2$
   $\qquad = (15.00)^2/(12.00)^2$
   $\qquad = 1.4038$
   conclusion:
   Do not reject H_o; there is not sufficient evidence to conclude that $\sigma_1^2 > \sigma_2^2$.

5. Let the traditional method be group 1.
   original claim: $\sigma_1^2 = \sigma_2^2$
   H_o: $\sigma_1^2 = \sigma_2^2$
   H_1: $\sigma_1^2 \neq \sigma_2^2$
   $\alpha = .05$
   C.R. $F < F_{29,.975}^{24} = [.4527]$
   $\qquad F > F_{29,.025}^{24} = 2.1540$
   calculations:
   $\qquad F = s_1^2/s_2^2$
   $\qquad\quad = (.37)^2/(.31)^2$
   $\qquad\quad = 1.4246$
   conclusion:

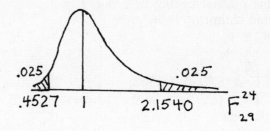

   Do not reject H_o; there is not sufficient evidence to conclude that $\sigma_1^2 \neq \sigma_2^2$.
   No; one cannot be 95% certain that there is really any difference in the variations.

7. Let the first time period be group 1.
   original claim: $\sigma_1^2 = \sigma_2^2$
   H_o: $\sigma_1^2 = \sigma_2^2$
   H_1: $\sigma_1^2 \neq \sigma_2^2$
   $\alpha = .05$
   C.R. $F < F_{11,.975}^{11} = [.2836]$
   $\qquad F > F_{11,.025}^{11} = 3.5257$
   calculations:
   $\qquad F = s_1^2/s_2^2$
   $\qquad\quad = (11.07)^2/(10.39)^2$
   $\qquad\quad = 1.1352$
   conclusion:
   Do not reject H_o; there is not sufficient evidence to conclude that $\sigma_1^2 \neq \sigma_2^2$.

9. Let TWA be group 1.
   original claim: $\sigma_1^2 = \sigma_2^2$
   H_o: $\sigma_1^2 = \sigma_2^2$
   H_1: $\sigma_1^2 \neq \sigma_2^2$
   $\alpha = .10$
   C.R. $F < F_{39,.95}^{34} = [.5581]$
   $\qquad F > F_{39,.05}^{34} = 1.7444$
   calculations:
   $\qquad F = s_1^2/s_2^2$
   $\qquad\quad = (3065)^2/(2960)^2$
   $\qquad\quad = 1.0722$
   conclusion:
   Do not reject H_o; there is not sufficient evidence to conclude that $\sigma_1^2 \neq \sigma_2^2$.

11. Let the manual keying be group 1.
   original claim: $\sigma_1^2 = \sigma_2^2$
   H_o: $\sigma_1^2 = \sigma_2^2$
   H_1: $\sigma_1^2 \neq \sigma_2^2$
   $\alpha = .02$
   C.R. $F < F_{9,.99}^{15} = [.2568]$
   $F > F_{9,.01}^{15} = 4.9621$
   calculations:
   $F = s_1^2/s_2^2$

   $ = 225.0/56.0$
   $ = 4.0179$
   conclusion:
   Do not reject H_o; there is not sufficient evidence to conclude that $\sigma_1^2 \neq \sigma_2^2$.

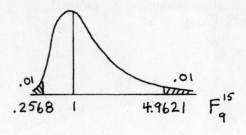

13. experimental$$control
   $n = 20 n = 16$
   $\Sigma x = 1327.61 \Sigma x = 1660.28$
   $\Sigma x^2 = 96477.3313 \Sigma x^2 = 181167.5442$
   $\bar{x} = 66.3805 \bar{x} = 103.7675$
   $s^2 = 439.469 s^2 = 592.296$
   Let the controls be group 1.
   original claim: $\sigma_1^2 = \sigma_2^2$
   H_o: $\sigma_1^2 = \sigma_2^2$
   H_1: $\sigma_1^2 \neq \sigma_2^2$
   $\alpha = .05$
   C.R. $F < F_{19,.975}^{15} = [.3629]$
   $F > F_{19,.025}^{15} = 2.6171$
   calculations:
   $F = s_1^2/s_2^2$

   $ = 592.296/439.469$
   $ = 1.3478$
   conclusion:
   Do not reject H_o; there is not sufficient evidence to conclude that $\sigma_1^2 \neq \sigma_2^2$.

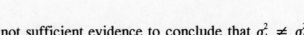

15. placebo$$calcium
   $n = 13 n = 15$
   $\Sigma x = 1490.5 \Sigma x = 1740.4$
   $\Sigma x^2 = 171965.47 \Sigma x^2 = 202936.92$
   $\bar{x} = 114.65 \bar{x} = 116.03$
   $s^2 = 89.493 s^2 = 71.722$
   Let the placebos be group 1.
   original claim: $\sigma_1^2 = \sigma_2^2$
   H_o: $\sigma_1^2 = \sigma_2^2$
   H_1: $\sigma_1^2 \neq \sigma_2^2$
   $\alpha = .05$
   C.R. $F < F_{14,.975}^{12} = [.3147]$
   $F > F_{14,.025}^{12} = 3.0502$
   calculations:
   $F = s_1^2/s_2^2$

   $ = 89.493/71.722$
   $ = 1.2478$
   conclusion:
   Do not reject H_o; there is not sufficient evidence to conclude that $\sigma_1^2 \neq \sigma_2^2$.

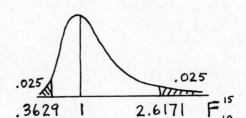

17. Let the first professor's scores be group 1.
    original claim: $\sigma_1^2 > \sigma_2^2$
    H_o: $\sigma_1^2 \leq \sigma_2^2$
    H_1: $\sigma_1^2 > \sigma_2^2$
    $\alpha = .05$
    C.R. F > $F_{24,.05}^{24} = 1.9838$

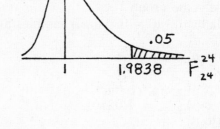

    calculations:
       $F = s_1^2/s_2^2$
          $= 103.4/39.7$
          $= 2.6045$
    conclusion:
       Reject H_o; there is sufficient evidence to conclude that $\sigma_1^2 > \sigma_2^2$.
    A student very weak in organic chemistry should choose the second professor -- i.e., the one
       whose grades have the smaller variance. The potentially extremely low scores (and -- to the
       disadvantage of very good students -- the potentially extremely high scores) will tend to turn
       out closer to the mean of the scores.

19. a. $F_L = F_{9,.975}^9 = 1/F_{9,.025}^9 = 1/4.0260 = .2484$

    $F_R = F_{9,.025}^9 = 4.0260$

    b. $F_L = F_{6,.975}^9 = 1/F_{9,.025}^6 = 1/4.3197 = .2315$

    $F_R = F_{6,.025}^9 = 5.5234$

    c. $F_L = F_{9,.975}^6 = 1/F_{6,.025}^9 = 1/5.5234 = .1810$

    $F_R = F_{9,.025}^6 = 4.3197$

    d. $F_L = F_{9,.99}^{24} = 1/F_{24,.025}^9 = 1/3.2560 = .3071$

    $F_R = F_{9,.01}^{24} = 4.7290$

    e. $F_L = F_{24,.99}^9 = 1/F_{9,.025}^{24} = 1/4.7290 = .2115$

    $F_R = F_{24,.01}^9 = 3.2560$

21. a. No. Adding a constant to each score does not affect the spread of the scores. All the
       standard deviations and variances remain the same, and so the F statistic (i.e., the ratio of
       the variances) is unchanged.
    b. No. Multiplying each score by a constant multiplies the standard deviation of those scores
       by that constant and the variance of those scores by the square of that constant. If this is
       done to both groups, so that each variance (i.e., the numerator and the denominator of the F
       statistic) is multiplied by the square of that constant, then the F statistic is unchanged.
    c. No. The change from Fahrenheit to Celsius is done by multiplication and addition. As
       noted in part (a), the additive constant does not affect the variances and, therefore, does not
       affect the F statistic. As noted in part (b), the multiplicative constant affects both the
       numerator and denominator of the F statistic in the same manner and, therefore, does not
       affect the value of the F statistic.

8-5 Inferences about Two Means: Independent and Small Samples

1. Let Sample A be group 1.
   original claim: $\mu_1 - \mu_2 = 0$ [small samples and σ unknown, first test H_o: $\sigma_1^2 = \sigma_2^2$]
   H_o: $\sigma_1^2 = \sigma_2^2$
   H_1: $\sigma_1^2 \neq \sigma_2^2$
   $\alpha = .05$
   C.R. $F < F_{9,.975}^9 = [.2484]$
        $F > F_{9,.025}^9 = 4.0260$

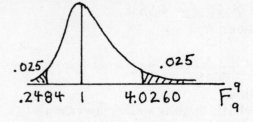

   calculations:
   $F = s_1^2/s_2^2$
     $= 50/25$
     $= 2.0000$
   conclusion:
        Do not reject H_o; there is not sufficient evidence to conclude that $\sigma_1^2 \neq \sigma_2^2$.
   Now proceed using s_p^2 for both s_1^2 and s_2^2
   $\bar{x}_1 - \bar{x}_2 = 200 - 185 = 15$
   $s_p^2 = (df_1 \cdot s_1^2 + df_2 \cdot s_2^2)/(df_1 + df_2)$
       $= (9 \cdot 50 + 9 \cdot 25)/(9 + 9)$
       $= 675/18$
       $= 37.5$
   H_o: $\mu_1 - \mu_2 = 0$
   H_1: $\mu_1 - \mu_2 \neq 0$
   $\alpha = .05$
   C.R. $t < -t_{18,.025} = -2.101$
        $t > t_{18,.025} = 2.101$

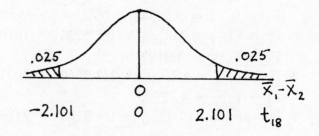

   calculations:
   $t_{\bar{x}_1 - \bar{x}_2} = (\bar{x}_1 - \bar{x}_2 - \mu_{\bar{x}_1 - \bar{x}_2})/s_{\bar{x}_1 - \bar{x}_2}$
       $= (15 - 0)/\sqrt{37.5/10 + 37.5/10}$
       $= 15/2.739$
       $= 5.477$
   conclusion:
        Reject H_o; there is sufficient evidence to conclude that $\mu_1 - \mu_2 \neq 0$ (in fact, $\mu_1 - \mu_2 > 0$).

3. Refer to exercise #1.
   $(\bar{x}_1 - \bar{x}_2) \pm t_{18,.025} \sqrt{s_p^2/n_1 + s_p^2/n_2}$
   $15 \pm 2.101 \cdot \sqrt{37.5/10 + 37.5/10}$
   15 ± 6
   $9 < \mu_1 - \mu_2 < 21$

5. Let the traditional method be group 1.
   original claim: $\mu_1 - \mu_2 = 0$ [small samples and σ unknown, first test H_o: $\sigma_1^2 = \sigma_2^2$]
   H_o: $\sigma_1^2 = \sigma_2^2$
   H_1: $\sigma_1^2 \neq \sigma_2^2$
   $\alpha = .05$
   C.R. $F < F_{29,.975}^{24} = [.4527]$
        $F > F_{29,.025}^{24} = 2.1540$

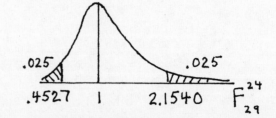

   calculations:
   $F = s_1^2/s_2^2$
     $= (.37)^2/(.31)^2$
     $= 1.4246$
   conclusion:
        Do not reject H_o; there is not sufficient evidence to conclude that $\sigma_1^2 \neq \sigma_2^2$.

Now proceed using s_p^2 for both s_1^2 and s_2^2

$\bar{x}_1 - \bar{x}_2 = 4.31 - 4.07 = .24$

$s_p^2 = (df_1 \cdot s_1^2 + df_2 \cdot s_2^2)/(df_1 + df_2)$

$\quad = (24 \cdot (.37)^2 + 29 \cdot (.31)^2)/(24 + 29)$

$\quad = 6.0725/53$

$\quad = .1146$

H_o: $\mu_1 - \mu_2 = 0$

H_1: $\mu_1 - \mu_2 \neq 0$

$\alpha = .05$

C.R. $t < -t_{53, .025} = -1.960$

$\quad t > t_{53, .025} = 1.960$

calculations:

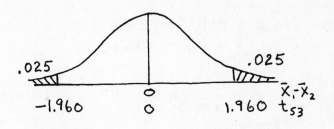

$t_{\bar{x}_1 - \bar{x}_2} = (\bar{x}_1 - \bar{x}_2 - \mu_{\bar{x}_1 - \bar{x}_2})/s_{\bar{x}_1 - \bar{x}_2}$

$\quad = (.24 - 0)/\sqrt{.1146/25 + .1146/30}$

$\quad = .24/.0917$

$\quad = 2.618$

conclusion:

Reject H_o; there is sufficient evidence to conclude that $\mu_1 - \mu_2 \neq 0$ (in fact, $\mu_1 - \mu_2 > 0$). A person buying a battery should prefer one manufactured by the traditional method.

7. Let the first time period be group 1.

original claim: $\mu_1 - \mu_2 = 0$ [small samples and σ unknown, first test H_o: $\sigma_1^2 = \sigma_2^2$]

H_o: $\sigma_1^2 = \sigma_2^2$

H_1: $\sigma_1^2 \neq \sigma_2^2$

$\alpha = .05$

C.R. $F < F_{11, .975}^{11} = [.2836]$

$\quad F > F_{11, .025}^{11} = 3.5257$

calculations:

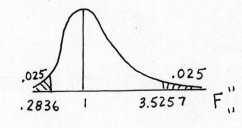

$F = s_1^2/s_2^2$

$\quad = (11.07)^2/(10.39)^2$

$\quad = 1.1352$

conclusion:

Do not reject H_o; there is not sufficient evidence to conclude that $\sigma_1^2 \neq \sigma_2^2$.

Now proceed using s_p^2 for both s_1^2 and s_2^2

$\bar{x}_1 - \bar{x}_2 = 46.42 - 51.00 = -4.58$

$s_p^2 = (df_1 \cdot s_1^2 + df_2 \cdot s_2^2)/(df_1 + df_2)$

$\quad = (11 \cdot (11.07)^2 + 11 \cdot (10.39)^2)/(11 + 11)$

$\quad = 2535.467/22$

$\quad = 115.25$

H_o: $\mu_1 - \mu_2 = 0$

H_1: $\mu_1 - \mu_2 \neq 0$

$\alpha = .05$

C.R. $t < -t_{22, .025} = -2.074$

$\quad t > t_{22, .025} = 2.074$

calculations:

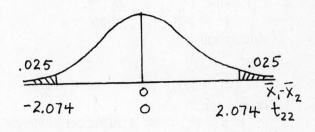

$t_{\bar{x}_1 - \bar{x}_2} = (\bar{x}_1 - \bar{x}_2 - \mu_{\bar{x}_1 - \bar{x}_2})/s_{\bar{x}_1 - \bar{x}_2}$

$\quad = (-4.58 - 0)/\sqrt{115.25/12 + 115.25/12}$

$\quad = -4.58/4.383$

$\quad = -1.045$

conclusion:

Do not reject H_o; there is not sufficient evidence to conclude that $\mu_1 - \mu_2 \neq 0$.

9. Let the manual keying be group 1.

original claim: $\mu_1 - \mu_2 = 0$ [small samples and σ unknown, first test H_o: $\sigma_1^2 = \sigma_2^2$]

H_o: $\sigma_1^2 = \sigma_2^2$

H_1: $\sigma_1^2 \neq \sigma_2^2$

$\alpha = .02$

C.R. F < $F_{9,.99}^{15}$ = [.2568]

F > $F_{9,.01}^{15}$ = 4.9621

calculations:

$F = s_1^2/s_2^2$

$= 225.0/56.0$

$= 4.0179$

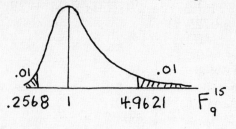

conclusion:

Do not reject H_o; there is not sufficient evidence to conclude that $\sigma_1^2 \neq \sigma_2^2$.

Now proceed using s_p^2 for both s_1^2 and s_2^2

$\bar{x}_1 - \bar{x}_2 = 157.6 - 112.4 = 45.2$

$s_p^2 = (df_1 \cdot s_1^2 + df_2 \cdot s_2^2)/(df_1 + df_2)$

$= (15 \cdot 225.0 + 9 \cdot 56.0)/(15 + 9)$

$= 3879/24$

$= 161.625$

H_o: $\mu_1 - \mu_2 = 0$

H_1: $\mu_1 - \mu_2 \neq 0$

$\alpha = .02$

C.R. t < $-t_{24,.01}$ = -2.492

t > $t_{24,.01}$ = 2.492

calculations:

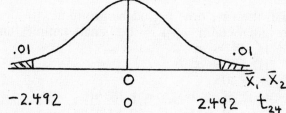

$t_{\bar{x}_1 - \bar{x}_2} = (\bar{x}_1 - \bar{x}_2 - \mu_{\bar{x}_1 - \bar{x}_2})/s_{\bar{x}_1 - \bar{x}_2}$

$= (45.2 - 0)/\sqrt{161.625/16 + 161.625/10}$

$= 45.2/5.125$

$= 8.820$

conclusion:

Reject H_o; there is sufficient evidence to conclude that $\mu_1 - \mu_2 \neq 0$ (in fact, $\mu_1 - \mu_2 > 0$).

11. Let the night shift be group 1.

confidence interval: $\mu_1 - \mu_2$ [small samples and σ unknown, first test H_o: $\sigma_1^2 = \sigma_2^2$]

H_o: $\sigma_1^2 = \sigma_2^2$

H_1: $\sigma_1^2 \neq \sigma_2^2$

$\alpha = .05$

C.R. F < $F_{24,.975}^{29}$ = [.4643]

F > $F_{24,.025}^{29}$ = 2.2090

calculations:

$F = s_1^2/s_2^2$

$= (.22)^2/(.14)^2$

$= 2.4694$

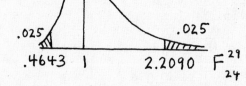

conclusion:

Reject H_o; there is sufficient evidence to conclude that $\sigma_1^2 \neq \sigma_2^2$ (in fact, $\sigma_1^2 > \sigma_2^2$).

Now proceed using s_1^2 and s_2^2 and the smaller degrees of freedom (viz., $df_2 = 24$).

$\bar{x}_1 - \bar{x}_2 = 11.85 - 12.02 = -.17$

$(\bar{x}_1 - \bar{x}_2) \pm t_{24,.025}\sqrt{s_1^2/n_1 + s_2^2/n_2}$

$-.17 \pm 2.064 \cdot \sqrt{(.22)^2/30 + (.14)^2/25}$

$-.17 \pm .10$

$-.27 < \mu_1 - \mu_2 < -.07$

13. Let the O-C patients be group 1.
    original claim: $\mu_1-\mu_2 = 0$ [small samples and σ unknown, first test H_o: $\sigma_1^2 = \sigma_2^2$]
    H_o: $\sigma_1^2 = \sigma_2^2$
    H_1: $\sigma_1^2 \neq \sigma_2^2$
    $\alpha = .05$ [or any other appropriate value]
    C.R. $F < F_{9,.975}^9 = [.2484]$
    $\quad\quad F > F_{9,.025}^9 = 4.0260$
    calculations:
    $\quad F = s_1^2/s_2^2$
    $\quad\quad = (.08)^2/(.08)^2$
    $\quad\quad = 1.0000$

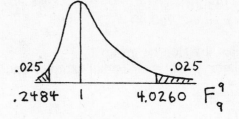

    conclusion:
        Do not reject H_o; there is not sufficient evidence to conclude that $\sigma_1^2 \neq \sigma_2^2$.
    Now proceed using s_p^2 for both s_1^2 and s_2^2
    $\bar{x}_1-\bar{x}_2 = .34 - .45 = -.11$
    $s_p^2 = (df_1 \cdot s_1^2 + df_2 \cdot s_2^2)/(df_1 + df_2)$
    $\quad\quad = (9 \cdot (.08)^2 + 9 \cdot (.08)^2)/(9 + 9)$
    $\quad\quad = .1152/18$
    $\quad\quad = .0064$
    H_o: $\mu_1-\mu_2 = 0$
    H_1: $\mu_1-\mu_2 \neq 0$
    $\alpha = .05$ [or any other appropriate value]
    C.R. $t < -t_{18,.025} = -2.101$
    $\quad\quad t > t_{18,.025} = 2.101$
    calculations:

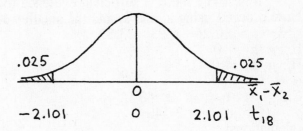

    $t_{\bar{x}_1-\bar{x}_2} = (\bar{x}_1-\bar{x}_2 - \mu_{\bar{x}_1-\bar{x}_2})/s_{\bar{x}_1-\bar{x}_2}$
    $\quad\quad = (-.11 - 0)/\sqrt{.0064/10 + .0064/10}$
    $\quad\quad = -.11/.0358$
    $\quad\quad = -3.075$
    conclusion:
    Reject H_o; there is sufficient evidence to conclude that $\mu_1-\mu_2 \neq 0$ (in fact, $\mu_1-\mu_2 < 0$).
    Based on this result, it does appear that obsessive-compulsive disorders have a biological
    indicator.

15. Let the O-C patients be group 1.
    confidence interval: $\mu_1-\mu_2$ [small samples and σ unknown, first test H_o: $\sigma_1^2 = \sigma_2^2$]
    H_o: $\sigma_1^2 = \sigma_2^2$
    H_1: $\sigma_1^2 \neq \sigma_2^2$
    $\alpha = .05$
    C.R. $F < F_{9,.975}^9 = [.2484]$
    $\quad\quad F > F_{9,.025}^9 = 4.0260$
    calculations:
    $\quad F = s_1^2/s_2^2$

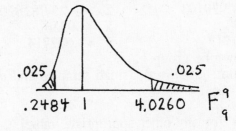

    $\quad\quad = (156.84)^2/(137.97)^2$
    $\quad\quad = 1.2922$
    conclusion:
        Do not reject H_o; there is not sufficient evidence to conclude that $\sigma_1^2 \neq \sigma_2^2$.
    Now proceed using s_p^2 for both s_1^2 and s_2^2
    $\bar{x}_1-\bar{x}_2 = 1390.03 - 1268.41 = 121.62$
    $s_p^2 = (df_1 \cdot s_1^2 + df_2 \cdot s_2^2)/(df_1 + df_2)$
    $\quad\quad = (9 \cdot (156.84)^2 + 9 \cdot (137.97)^2)/(9 + 9)$
    $\quad\quad = 43634.5065/18$
    $\quad\quad = 21817.3$

$$(\bar{x}_1 - \bar{x}_2) \pm t_{18,.025}\sqrt{s_p^2/n_1 + s_p^2/n_2}$$

121.62 $\pm$ 2.101 $\cdot \sqrt{21817.3/10 + 21817.3/10}$

121.62 $\pm$ 138.78

$-17.16 < \mu_1 - \mu_2 < 260.40$

17. Let the sales division be group 1.
   original claim: $\mu_1 - \mu_2 = 0$ [small samples and σ unknown, first test H_o: $\sigma_1^2 = \sigma_2^2$]
   H_o: $\sigma_1^2 = \sigma_2^2$
   H_1: $\sigma_1^2 \neq \sigma_2^2$
   $\alpha = .02$
   C.R. $F < F_{19,.99}^{39} = [.4221]$
   $\quad\ F > F_{19,.01}^{39} = 2.7608$
   calculations:
   $\quad\ F = s_1^2/s_2^2$
   $\quad\quad = (8.65)^2/(4.93)^2$
   $\quad\quad = 3.0785$

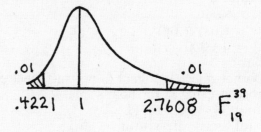

   conclusion:
   $\quad$ Reject H_o; there is sufficient evidence to conclude that $\sigma_1^2 \neq \sigma_2^2$ (in fact, $\sigma_1^2 > \sigma_2^2$).
   Now proceed using s_1^2 and s_2^2 and the smaller degrees of freedom (viz., $df_2 = 19$).
   $\bar{x}_1 - \bar{x}_2 = 10.26 - 6.93 = 3.33$
   H_o: $\mu_1 - \mu_2 = 0$
   H_1: $\mu_1 - \mu_2 \neq 0$
   $\alpha = .02$
   C.R. $t < -t_{19,.01} = -2.540$
   $\quad\ t > t_{19,.01} = 2.540$
   calculations:
   $\quad t_{\bar{x}_1-\bar{x}_2} = (\bar{x}_1 - \bar{x}_2 - \mu_{\bar{x}_1-\bar{x}_2})/s_{\bar{x}_1-\bar{x}_2}$
   $\quad\quad = (3.33 - 0)/\sqrt{(8.65)^2/40 + (4.93)^2/20}$
   $\quad\quad = 3.33/1.757$
   $\quad\quad = 1.896$

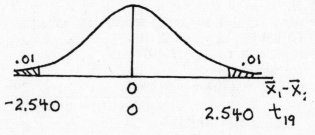

   conclusion:
   $\quad$ Do not reject H_o; there is not sufficient evidence to conclude that $\mu_1 - \mu_2 \neq 0$.

19. red
   $\quad$ n = 17
   $\quad \Sigma x = 15.373$
   $\quad \Sigma x^2 = 13.929193$
   $\quad \bar{x} = .9043$
   $\quad s^2 = .001717$

   brown
   $\quad$ n = 30
   $\quad \Sigma x = 27.769$
   $\quad \Sigma x^2 = 25.782620$
   $\quad \bar{x} = .9256$
   $\quad s^2 = .002714$

   Let the brown be group 1.
   original claim: $\mu_1 - \mu_2 = 0$ [small samples and σ unknown, first test H_o: $\sigma_1^2 = \sigma_2^2$]
   H_o: $\sigma_1^2 = \sigma_2^2$
   H_1: $\sigma_1^2 \neq \sigma_2^2$
   $\alpha = .05$ [or any other appropriate value]
   C.R. $F < F_{16,.975}^{29} = [.4301]$
   $\quad\ F > F_{16,.025}^{29} = 2.5678$
   calculations:
   $\quad\ F = s_1^2/s_2^2$
   $\quad\quad = .002714/.001717$
   $\quad\quad = 1.5802$

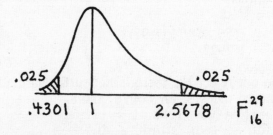

   conclusion:
   $\quad$ Do not reject H_o; there is not sufficient evidence to conclude that $\sigma_1^2 \neq \sigma_2^2$.

Now proceed using s_p^2 for both s_1^2 and s_2^2

$\bar{x}_1 - \bar{x}_2 = .9256 - .9043 = .0213$

$s_p^2 = (df_1 \cdot s_1^2 + df_2 \cdot s_2^2)/(df_1 + df_2)$

$\quad = (29 \cdot (.002714) + 16 \cdot (.001717))/(29 + 16)$

$\quad = .106187/45$

$\quad = .002360$

H_o: $\mu_1 - \mu_2 = 0$

H_1: $\mu_1 - \mu_2 \neq 0$

$\alpha = .05$ [or any other appropriate value]

C.R. $t < -t_{45, .025} = -1.960$

$\quad\quad t > t_{45, .025} = 1.960$

calculations:

$t_{\bar{x}_1 - \bar{x}_2} = (\bar{x}_1 - \bar{x}_2 - \mu_{\bar{x}_1 - \bar{x}_2})/s_{\bar{x}_1 - \bar{x}_2}$

$\quad = (.0213 - 0)/\sqrt{.002360/30 + .002360/17}$

$\quad = .0213/.0147$

$\quad = 1.447$

conclusion:

Do not reject H_o; there is not sufficient evidence to conclude that $\mu_1 - \mu_2 \neq 0$.

No. One cannot be 95% certain that the differences observed reflect anything other than the normal variability expected to occur when the process is in control as specified.

21. $A = s_1^2/n_1 = (.22)^2/30 = .001613$

$\quad B = s_2^2/n_2 = (.14)^2/25 = .000784$

$\quad df = (A + B)^2/(A^2/df_1 + B^2/df_2)$

$\quad\quad = (.002397)^2/(.000002602/29 + .000000614/24)$

$\quad\quad = .000005745/.000000115$

$\quad\quad = 49.8$

$(\bar{x}_1 - \bar{x}_2) \pm t_{49.8, .025}\sqrt{s_1^2/n_1 + s_2^2/n_2}$

$-.17 \pm 1.960 \cdot \sqrt{(.22)^2/30 + (.14)^2/25}$

$-.17 \pm .10$

$-.27 < \mu_1 - \mu_2 < -.07$

This is the same result obtained in exercise #11.

23. a. No really. If $s_1 = s_2$, then $F = s_1^2/s_2^2 = 1.0000$. Since all upper tail F critical values are greater than 1.0000 and all lower tail F critical values are less than 1.0000, the value $F = 1.0000$ will not be in the critical region for any df_1 and df_2.

b. Since s_p^2 is a weighted average of s_1^2 and s_2^2, $s_1 = s_2 = s$ means that $s_p^2 = s^2$.

8-6 Inferences about Two Proportions

NOTE: To be consistent with the notation of the previous sections, and thereby reinforcing the patterns and concepts presented in those sections, the manual uses the "usual" z formula written to apply to $\hat{p}_1$-$\hat{p}_2$'s

$$z_{\hat{p}_1-\hat{p}_2} = (\hat{p}_1-\hat{p}_2 - \mu_{\hat{p}_1-\hat{p}_2})/\sigma_{\hat{p}_1-\hat{p}_2}$$

with $\mu_{\hat{p}_1-\hat{p}_2} = p_1 - p_2$

and $\sigma_{\hat{p}_1-\hat{p}_2} = \sqrt{\overline{pq}/n_1 + \overline{pq}/n_2}$ [when H_o includes $p_1=p_2$]

where $\bar{p} = (x_1 + x_2)/(n_1 + n_2)$

And so the formula for the z statistic may also be written as

$$z_{\hat{p}_1-\hat{p}_2} = ((\hat{p}_1-\hat{p}_2) - (p_1-p_2))/\sqrt{\overline{pq}/n_1 + \overline{pq}/n_2}$$

1. $\hat{p}_1 = x_1/n_1 = 45/100 = .450$
   $\hat{p}_2 = x_2/n_2 = 115/200 = .575$
   $\hat{p}_1-\hat{p}_2 = .450 - .575 = -.125$
   a. $\bar{p} = (x_1 + x_2)/(n_1 + n_2) = (45 + 115)/(100 + 200) = 160/300 = .533$
   b. $z_{\hat{p}_1-\hat{p}_2} = (\hat{p}_1-\hat{p}_2 - \mu_{\hat{p}_1-\hat{p}_2})/\sigma_{\hat{p}_1-\hat{p}_2}$
      $= (-.125 - 0)/\sqrt{(.533)(.467)/100 + (.533)(.467)/200}$
      $= -.125/.0611$
      $= -2.046$
   c. $\pm z_{.025} = \pm 1.960$
   d. P-value $= 2 \cdot P(z < -2.05)$
      $= 2 \cdot (.5000 - .4798)$
      $= 2 \cdot (.0202)$
      $= .0404$

NOTE: Since $\bar{p}$ is the weighted average of $\hat{p}_1$ and $\hat{p}_2$, it must always fall between those two values. If it does not, then an error has been made that must be corrected before proceeding. Calculation of $\sigma_{\hat{p}_1-\hat{p}_2} = \sqrt{\overline{pq}/n_1 + \overline{pq}/n_2}$ can be accomplished with no round-off loss on most calculators by calculating $\bar{p}$ and proceeding as follows: STORE 1-RECALL = * RECALL = STORE RECALL $\div$ n_1 + RECALL $\div$ n_2 = $\sqrt{}$. The quantity $\sigma_{\hat{p}_1-\hat{p}_2}$ may then be STORED for future use. Each calculator is different -- learn how your calculator works, and do the homework on the same calculator you will use for the exam. If you have any questions about performing/storing calculations on your calculator, check with your instructor or class assistant.

3. Let the Democrats be group 1.
   original claim: $p_1-p_2 = 0$
   $\hat{p}_1 = x_1/n_1 = .35$ [see NOTE below]
   $\hat{p}_2 = x_2/n_2 = .41$ [see NOTE below]
   $\hat{p}_1-\hat{p}_2 = .35 - .41 = -.06$
   $\bar{p} = (x_1 + x_2)/(n_1 + n_2) = (.35 \cdot 552 + .41 \cdot 417)/(552 + 417) = 364.17/969 = .376$
   H_o: $p_1-p_2 = 0$
   H_1: $p_1-p_2 \neq 0$
   $\alpha = .05$
   C.R. $z < -z_{.025} = -1.960$
   $\quad\quad z > z_{.025} = 1.960$
   calculations:

   $z_{\hat{p}_1-\hat{p}_2} = (\hat{p}_1-\hat{p}_2 - \mu_{\hat{p}_1-\hat{p}_2})/\sigma_{\hat{p}_1-\hat{p}_2}$
   $= (-.06 - 0)/\sqrt{(.376)(.624)/552 + (.376)(.624)/417}$
   $= -.06/.0314$
   $= -1.909$
   conclusion:
   Do not reject H_o; there is not sufficient evidence to conclude that $p_1-p_2 \neq 0$.

NOTE: In the preceding problem x_1 and x_2 were not given and must be deduced from the values of $\hat{p}_1 = 35\%$ and $\hat{p}_2 = 41\%$. Unfortunately any x_1 between 191 and 195 inclusive and any x_2 between 169 and 173 inclusive produces the given percents. Whenever this occurs (i.e., x_1 and x_2 cannot be determined exactly), the estimate $x = \hat{p} \cdot n$ will be employed without further manipulation -- even if that estimate is not a whole number.

5. Let those not wearing seat belts be group 1.
   original claim: $p_1 - p_2 > 0$
   $\hat{p}_1 = x_1/n_1 = 50/290 = .1724$
   $\hat{p}_2 = x_2/n_2 = 16/123 = .1301$
   $\hat{p}_1 - \hat{p}_2 = .1724 - .1301 = .0423$
   $\bar{p} = (x_1 + x_2)/(n_1 + n_2) = (50 + 16)/(290 + 123) = 66/413 = .160$
   H_o: $p_1 - p_2 \le 0$
   H_1: $p_1 - p_2 > 0$
   $\alpha = .05$
   C.R. $z > z_{.05} = 1.645$
   calculations:

   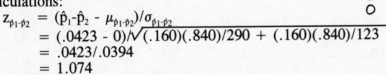

   $z_{\hat{p}1-\hat{p}2} = (\hat{p}_1 - \hat{p}_2 - \mu_{\hat{p}1-\hat{p}2})/\sigma_{\hat{p}1-\hat{p}2}$
   $\quad = (.0423 - 0)/\sqrt{(.160)(.840)/290 + (.160)(.840)/123}$
   $\quad = .0423/.0394$
   $\quad = 1.074$
   conclusion:
   Do not reject H_o; there is not sufficient evidence to conclude that $p_1 - p_2 > 0$.

7. Let the 18-24 year olds be group 1.
   $\hat{p}_1 = x_1/n_1 = .360$
   $\hat{p}_2 = x_2/n_2 = .540$
   $\hat{p}_1 - \hat{p}_2 = .360 - .540 = -.180$
   $(\hat{p}_1 - \hat{p}_2) \pm z_{.025}\sqrt{\hat{p}_1\hat{q}_1/n_1 + \hat{p}_2\hat{q}_2/n_2}$
   $-.180 \pm 1.960 \cdot \sqrt{(.360)(.640)/200 + (.540)(.460)/250}$
   $-.180 \pm .091$
   $-.271 < p_1 - p_2 < -.089$

9. Let the vinyl gloves be group 1.
   original claim: $p_1 - p_2 > 0$
   $\hat{p}_1 = x_1/n_1 = .63$
   $\hat{p}_2 = x_2/n_2 = .07$
   $\hat{p}_1 - \hat{p}_2 = .63 - .07 = .56$
   $\bar{p} = (x_1 + x_2)/(n_1 + n_2) = (.63 \cdot 240 + .07 \cdot 240)/(240 + 240) = 168/480 = .350$
   H_o: $p_1 - p_2 \le 0$
   H_1: $p_1 - p_2 > 0$
   $\alpha = .005$
   C.R. $z > z_{.005} = 2.575$
   calculations:
   $z_{\hat{p}1-\hat{p}2} = (\hat{p}_1 - \hat{p}_2 - \mu_{\hat{p}1-\hat{p}2})/\sigma_{\hat{p}1-\hat{p}2}$
   $\quad = (.56 - 0)/\sqrt{(.350)(.650)/240 + (.350)(.650)/240}$
   $\quad = .56/.0435$
   $\quad = 12.861$
   conclusion:
   Reject H_o; there is sufficient evidence to conclude that $p_1 - p_2 > 0$.

11. Let the central city be group 1.
    original claim: $p_1 - p_2 = 0$
    $\hat{p}_1 = x_1/n_1 = .289$
    $\hat{p}_2 = x_2/n_2 = .171$
    $\hat{p}_1 - \hat{p}_2 = .289 - .171 = .118$
    $\bar{p} = (x_1 + x_2)/(n_1 + n_2) = (.289 \cdot 294 + .171 \cdot 1015)/(294 + 1015) = 258.5/1309 = .198$
    H_o: $p_1 - p_2 = 0$
    H_1: $p_1 - p_2 \neq 0$
    $\alpha = .01$
    C.R. $z < -z_{.005} = -2.575$
    $\quad z > z_{.005} = 2.575$
    calculations:

$$z_{\hat{p}_1 - \hat{p}_2} = (\hat{p}_1 - \hat{p}_2 - \mu_{\hat{p}_1 - \hat{p}_2})/\sigma_{\hat{p}_1 - \hat{p}_2}$$
$$= (.118 - 0)/\sqrt{(.198)(.802)/294 + (.198)(.802)/1015}$$
$$= .118/.0264$$
$$= 4.475$$

conclusion:
    Reject H_o; there is sufficient evidence to conclude that $p_1 - p_2 \neq 0$ (in fact, $p_1 - p_2 > 0$).

13. Let the public colleges be group 1.
    original claim: $p_1 - p_2 = 0$
    $\hat{p}_1 = x_1/n_1 = .30$
    $\hat{p}_2 = x_2/n_2 = .26$
    $\hat{p}_1 - \hat{p}_2 = .30 - .26 = .04$
    $\bar{p} = (x_1 + x_2)/(n_1 + n_2) = (.30 \cdot 1000 + .26 \cdot 500)/(1000 + 500) = 430/1500 = .287$
    H_o: $p_1 - p_2 = 0$
    H_1: $p_1 - p_2 \neq 0$
    $\alpha = .05$
    C.R. $z < -z_{.025} = -1.960$
    $\quad z > z_{.025} = 1.960$
    calculations:

$$z_{\hat{p}_1 - \hat{p}_2} = (\hat{p}_1 - \hat{p}_2 - \mu_{\hat{p}_1 - \hat{p}_2})/\sigma_{\hat{p}_1 - \hat{p}_2}$$
$$= (.04 - 0)/\sqrt{(.287)(.713)/1000 + (.287)(.713)/500}$$
$$= .04/.0248$$
$$= 1.615$$

conclusion:
    Do not reject H_o; there is not sufficient evidence to conclude that $p_1 - p_2 \neq 0$.

15. Let the males be group 1.
    original claim: p_1-$p_2 = 0$
    $\hat{p}_1 = x_1/n_1 = 25/86 = .29070$
    $\hat{p}_2 = x_2/n_2 = 11/39 = .28205$
    $\hat{p}_1$-$\hat{p}_2 = .29070 - .28205 = .00865$
    $\bar{p} = (x_1 + x_2)/(n_1 + n_2) = (25 + 11)/(86 + 39) = 36/125 = .288$
    H_o: p_1-$p_2 = 0$
    H_1: p_1-$p_2 \neq 0$
    $\alpha = .05$
    C.R. $z < -z_{.025} = -1.960$
    $z > z_{.025} = 1.960$
    calculations:
    $z_{\hat{p}_1-\hat{p}_2} = (\hat{p}_1-\hat{p}_2 - \mu_{\hat{p}_1-\hat{p}_2})/\sigma_{\hat{p}_1-\hat{p}_2}$
    $= (.00865 - 0)/\sqrt{(.288)(.712)/86 + (.288)(.712)/39}$
    $= .0865/.0874$
    $= .099$
    conclusion:
    Do not reject H_o; there is not sufficient evidence to conclude that p_1-$p_2 \neq 0$.

17. Let the Californians be group 1.
    original claim: p_1-$p_2 = .25$
    $\hat{p}_1 = x_1/n_1 = 210/500 = .42$
    $\hat{p}_2 = x_2/n_2 = 120/500 = .24$
    $\hat{p}_1$-$\hat{p}_2 = .42 - .24 = .18$
    H_o: p_1-$p_2 = .25$
    H_1: p_1-$p_2 \neq .25$
    $\alpha = .05$
    C.R. $z < -z_{.025} = -1.960$
    $z > z_{.025} = 1.960$
    calculations:
    $z_{\hat{p}_1-\hat{p}_2} = (\hat{p}_1-\hat{p}_2 - \mu_{\hat{p}_1-\hat{p}_2})/\sigma_{\hat{p}_1-\hat{p}_2}$
    $= (.42 - .25)/\sqrt{(.42)(.58)/500 + (.24)(.76)/500}$
    $= -.07/.0292$
    $= -2.398$
    conclusion:
    Reject H_o; there is sufficient evidence to conclude that p_1-$p_2 \neq .25$ (in fact, p_1-$p_2 < .25$).

19. $E^2 = (z_{\alpha/2})^2(p_1q_1/n_1 + p_2q_2/n_2)$ — squaring the original equation
    $E^2 = (z_{\alpha/2})^2(.25/n_1 + .25/n_2)$ — setting the unknown proportions to .5
    $E^2 = (z_{\alpha/2})^2(.25/n + .25/n)$ — requiring $n_1 = n_2 = n$
    $E^2 = (z_{\alpha/2})^2(.50/n)$ — addition
    $n = (z_{\alpha/2})^2(.50)/E^2$ — solving for n
    $= (1.960)^2(.50)/(.03)^2$ — for $\alpha = .05$ and $E = .03$
    $= 2134.2$ rounded up to 2135

Review Exercises

1. Let Orange County be group 1.
   concerns $\mu_1 - \mu_2$ [$n_1 > 30$ and $n_2 > 30$, use z (with s's for σ's)]
   $\bar{x}_1 - \bar{x}_2 = 183.0 - 253.1 = -70.1$

   a. original claim: $\mu_1 - \mu_2 = 0$
      H_o: $\mu_1 - \mu_2 = 0$
      H_1: $\mu_1 - \mu_2 \neq 0$
      $\alpha = .05$
      C.R. $z < -z_{.025} = -1.960$
      $\quad\;\; z > z_{.025} = 1.960$
      calculations:

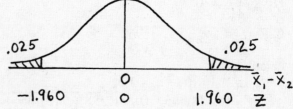

$$z_{\bar{x}_1 - \bar{x}_2} = (\bar{x}_1 - \bar{x}_2 - \mu_{\bar{x}_1 - \bar{x}_2})/\sigma_{\bar{x}_1 - \bar{x}_2}$$
$$= (-70.1 - 0)/\sqrt{(21.0)^2/40 + (29.2)^2/50}$$
$$= -70.1/5.299$$
$$= -13.229$$

   conclusion:
       Reject H_o; there is sufficient evidence to conclude that $\mu_1 - \mu_2 \neq 0$ (in fact, $\mu_1 - \mu_2 < 0$).

   b. $(\bar{x}_1 - \bar{x}_2) \pm z_{.025}\sqrt{\sigma_1^2/n_1 + \sigma_2^2/n_2}$
   $-70.1 \pm 1.960 \cdot \sqrt{(21.0)^2/40 + (29.2)^2/50}$
   -70.1 ± 10.4
   $-80.5 < \mu_1 - \mu_2 < -59.7$

3. original claim $\mu_d > 0$ [$n \leq 30$ and σ_d unknown, use t]
   $d = x_{new} - x_{ord}$: 2.1 2.1 .5 -.1 -.6 3.2 .4 1.5 1.8
   $n = 9$
   $\Sigma d = 10.9 \qquad\qquad \bar{d} = 1.21$
   $\Sigma d^2 = 25.33 \qquad\quad s_d = 1.231$
   H_o: $\mu_d \leq 0$
   H_1: $\mu_d > 0$
   $\alpha = .05$
   C.R. $t > t_{8,.05} = 1.860$
   calculations:
   $$t_{\bar{d}} = (\bar{d} - \mu_{\bar{d}})/s_{\bar{d}}$$
   $$= (1.21 - 0)/(1.231/\sqrt{9})$$
   $$= 1.21/.410$$
   $$= 2.951$$
   conclusion:
       Reject H_o; there is sufficient evidence to conclude that $\mu_d > 0$.
   No; the results do not give the postal service evidence of fraud, but they do give evidence that
   Minton's claim is valid.

5. Let Albany County be group 1.
   original claim: $p_1-p_2 > 0$ [concerns p's, use z]
   $\hat{p}_1 = x_1/n_1 = 558/24{,}384 = .02288$
   $\hat{p}_2 = x_2/n_2 = 1214/166{,}197 = .00730$
   $\hat{p}_1-\hat{p}_2 = .02288 - .00730 = .01558$
   $\bar{p} = (x_1 + x_2)/(n_1 + n_2) = (558 + 1214)/(24{,}384 + 166{,}197) = 1772/190{,}581 = .00930$
   H_o: $p_1-p_2 \leq 0$
   H_1: $p_1-p_2 > 0$
   $\alpha = .01$
   C.R. $z > z_{.01} = 2.327$
   calculations:

   $z_{\hat{p}_1-\hat{p}_2} = (\hat{p}_1-\hat{p}_2 - \mu_{\hat{p}_1-\hat{p}_2})/\sigma_{\hat{p}_1-\hat{p}_2}$
   $= (.01558 - 0)/\sqrt{(.00930)(.99070)/24{,}384 + (.00930)(.99070)/166{,}197}$
   $= .01558/.0006582$
   $= 23.671$
   conclusion:
   Reject H_o; there is sufficient evidence to conclude that $p_1-p_2 > 0$.

7. Let the women be group 1.
   original claim: $\mu_1-\mu_2 = 0$ [$n_1 > 30$ and $n_2 > 30$, use z (with s's for σ's)]
   $\bar{x}_1-\bar{x}_2 = 538.82 - 525.23 = 13.59$
   H_o: $\mu_1-\mu_2 = 0$
   H_1: $\mu_1-\mu_2 \neq 0$
   $\alpha = .02$
   C.R. $z < -z_{.01} = -2.327$
       $z > z_{.01} = 2.327$
   calculations:
   $z_{\bar{x}_1-\bar{x}_2} = (\bar{x}_1-\bar{x}_2 - \mu_{\bar{x}_1-\bar{x}_2})/\sigma_{\bar{x}_1-\bar{x}_2}$
   $= (13.59 - 0)/\sqrt{(114.16)^2/68 + (97.23)^2/86}$
   $= 13.59/17.366$
   $= .783$
   conclusion:
   Do not reject H_o; there is not sufficient evidence to conclude that $\mu_1-\mu_2 \neq 0$.

9. Let the six-week program be group 1.
   original claim: $\mu_1-\mu_2 > 0$ [$n_1 > 30$ and $n_2 > 30$, use z (with s's for σ's)]
   $\bar{x}_1-\bar{x}_2 = 83.5 - 79.8 = 3.7$
   H_o: $\mu_1-\mu_2 \leq 0$
   H_1: $\mu_1-\mu_2 > 0$
   $\alpha = .01$
   C.R. $z > z_{.01} = 2.327$
   calculations:

   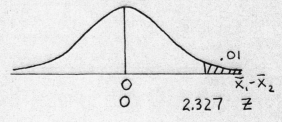

   $z_{\bar{x}_1-\bar{x}_2} = (\bar{x}_1-\bar{x}_2 - \mu_{\bar{x}_1-\bar{x}_2})/\sigma_{\bar{x}_1-\bar{x}_2}$
   $= (3.7 - 0)/\sqrt{(16.3)^2/60 + (19.2)^2/35}$
   $= 3.7/3.868$
   $= .957$
   conclusion:
   Do not reject H_o; there is not sufficient evidence to conclude that $\mu_1-\mu_2 > 0$.

11. Let the prepared students be group 1.
    original claim: $p_1-p_2 > 0$ [concerns p's, use z]
    $\hat{p}_1 = x_1/n_1 = 62/80 = .775$
    $\hat{p}_2 = x_2/n_2 = 23/50 = .460$
    $\hat{p}_1-\hat{p}_2 = .775 - .460 = .315$
    $\bar{p} = (x_1 + x_2)/(n_1 + n_2) = (62 + 23)/(80 + 50) = 85/130 = .654$
    H_o: $p_1-p_2 \leq 0$
    H_1: $p_1-p_2 > 0$
    $\alpha = .05$
    C.R. $z > z_{.05} = 1.645$
    calculations:

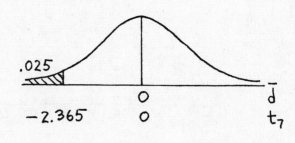

$z_{\hat{p}1-\hat{p}2} = (\hat{p}_1-\hat{p}_2 - \mu_{\hat{p}1-\hat{p}2})/\sigma_{\hat{p}1-\hat{p}2}$
$= (.315 - 0)/\sqrt{(.654)(.346)/80 + (.654)(.346)/50}$
$= .315/.0858$
$= 3.673$

    conclusion:
    Reject H_o; there is sufficient evidence to conclude that $p_1-p_2 > 0$.

13. a. original claim $\mu_d < 0$ [$n \leq 30$ and σ_d unknown, use t]
    $d = x_{bef} - x_{aft}$: -3 -2 -3 -3 -3 -3 -3 -3
    $n = 8$
    $\Sigma d = -23$ $\bar{d} = -2.875$
    $\Sigma d^2 = 67$ $s_d = .354$
    H_o: $\mu_d \geq 0$
    H_1: $\mu_d < 0$
    $\alpha = .025$
    C.R. $t < -t_{7,.025} = -2.365$
    calculations:
    $t_{\bar{d}} = (\bar{d} - \mu_{\bar{d}})/s_{\bar{d}}$
    $= (-2.875 - 0)/(.354/\sqrt{8})$
    $= -2.875/.125$
    $= -23.000$
    conclusion:
    Reject H_o; there is sufficient evidence to conclude that $\mu_d < 0$.

    b. $\bar{d} \pm t_{7,.025} \cdot s_d/\sqrt{n}$
    $-2.9 \pm 2.365 \cdot (.125)/\sqrt{8}$
    $-2.9 \pm .3$
    $-3.2 < \mu_d < -2.6$

15. Let Barrington be group 1.
   original claim: $\mu_1 - \mu_2 = 0$ [small samples and σ unknown, first test H_o: $\sigma_1^2 = \sigma_2^2$]
   H_o: $\sigma_1^2 = \sigma_2^2$
   H_1: $\sigma_1^2 \neq \sigma_2^2$
   $\alpha = .05$
   C.R. $F < F_{17,.975}^{23} = [.4054]$
   $\quad\ F > F_{17,.025}^{23} = 2.5598$
   calculations:
   $\quad F = s_1^2/s_2^2$
   $\quad\quad = (6.1)^2/(5.8)^2$
   $\quad\quad = 1.1061$

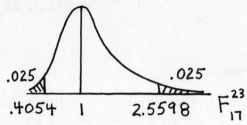

   conclusion:
   $\quad$ Do not reject H_o; there is not sufficient evidence to conclude that $\sigma_1^2 \neq \sigma_2^2$.
   Now proceed using s_p^2 for both s_1^2 and s_2^2
   $\overline{x}_1 - \overline{x}_2 = 80.6 - 85.7 = -5.1$
   $s_p^2 = (df_1 \cdot s_1^2 + df_2 \cdot s_2^2)/(df_1 + df_2)$
   $\quad\ = (23 \cdot (6.1)^2 + 17 \cdot (5.8)^2)/(23 + 17)$
   $\quad\ = 1427.71/40$
   $\quad\ = 35.69$
   H_o: $\mu_1 - \mu_2 = 0$
   H_1: $\mu_1 - \mu_2 \neq 0$
   $\alpha = .05$
   C.R. $t < -t_{40,.025} = -1.960$
   $\quad\ t > t_{40,.025} = 1.960$
   calculations:

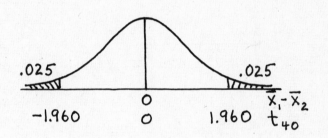

   $t_{\overline{x}_1 - \overline{x}_2} = (\overline{x}_1 - \overline{x}_2 - \mu_{\overline{x}_1 - \overline{x}_2})/s_{\overline{x}_1 - \overline{x}_2}$
   $\quad\quad = (-5.1 - 0)/\sqrt{35.69/24 + 35.69/18}$
   $\quad\quad = -5.1/1.863$
   $\quad\quad = -2.738$
   conclusion:
   $\quad$ Reject H_o; there is sufficient evidence to conclude that $\mu_1 - \mu_2 \neq 0$ (in fact, $\mu_1 - \mu_2 < 0$).

Chapter 9

Correlation and Regression

9-2 Correlation

1. The critical values below are taken from Table A-6.
   a. CV = ±.444; r = .502 indicates a significant (positive) linear correlation
   b. CV = ±.444; r = .203 indicates no significant linear correlation
   c. CV = ±.279; r = -.281 indicates a significant (negative) linear correlation

3. a.

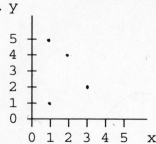

x	y	xy	x^2	y^2
1	1	1	1	1
1	5	5	1	25
2	4	8	4	16
3	2	6	9	4
7	12	20	15	46

   b. n = 4
   $\Sigma x = 7$
   $\Sigma x^2 = 15$
   $(\Sigma x)^2 = (7)^2 = 49$
   $\Sigma xy = 20$

   c. $n(\Sigma xy) - (\Sigma x)(\Sigma y) = 4(20) - (7)(12) = -4$
   $n(\Sigma x^2) - (\Sigma x)^2 = 4(15) - (7)^2 = 11$
   $n(\Sigma y^2) - (\Sigma y)^2 = 4(46) - (12)^2 = 40$
   $r = [n(\Sigma xy) - (\Sigma x)(\Sigma y)]/[\sqrt{n(\Sigma x^2) - (\Sigma x)^2} \cdot \sqrt{n(\Sigma y^2) - (\Sigma y)^2}]$
   $= [-4]/[\sqrt{11} \cdot \sqrt{40}]$
   $= -.191$

NOTE: In each of problems 5-20, the first variable listed is given the designation x, and the second variable listed is given the designation y. In correlation problems, the designation of x and y is arbitrary -- so long as a person remains consistent after making the designation. For part (d) of each problem, the following summary statistics should be saved: n, Σx, Σy, Σx^2, Σy^2, Σxy.

5. a.

   n = 8
   $\Sigma x = 62.28$
   $\Sigma y = 26$
   $\Sigma xy = 221.83$
   $\Sigma x^2 = 533.6532$
   $\Sigma y^2 = 104$

150

b. $n(\Sigma xy) - (\Sigma x)(\Sigma y) = 8(221.83) - (62.28)(26) = 155.36$
   $n(\Sigma x^2) - (\Sigma x)^2 = 8(533.6532) - (62.28)^2 = 390.4272$
   $n(\Sigma y^2) - (\Sigma y)^2 = 8(104) - (26)^2 = 156$
   $r = [n(\Sigma xy) - (\Sigma x)(\Sigma y)]/[\sqrt{n(\Sigma x^2) - (\Sigma x)^2} \cdot \sqrt{n(\Sigma y^2) - (\Sigma y)^2}]$
   $= 155.36/[\sqrt{390.4272} \cdot \sqrt{156}]$
   $= .630$

c. $H_o: \rho = 0$
   $H_1: \rho \neq 0$
   $\alpha = .05$
   C.R. $r < -.707$ <u>OR</u> C.R. $t < -t_{6,.025} = -2.447$
          $r > .707$ $t > t_{6,.025} = 2.447$
   calculations: calculations:
       $r = .630$ $t_r = (r - \mu_r)/s_r$
                              $= (.630 - 0)/\sqrt{(1-(.630)^2)/6}$
                              $= .630/.317$
                              $= 1.985$

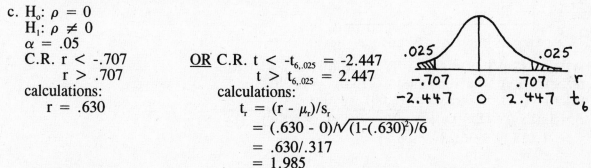

conclusion:
    Do not reject H_o; there is not sufficient evidence to conclude that $\rho \neq 0$.

7. a.

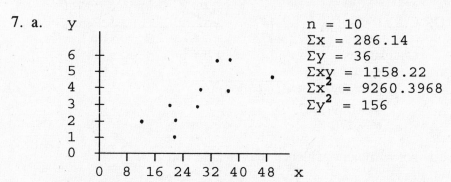

                                 $n = 10$
                                 $\Sigma x = 286.14$
                                 $\Sigma y = 36$
                                 $\Sigma xy = 1158.22$
                                 $\Sigma x^2 = 9260.3968$
                                 $\Sigma y^2 = 156$

b. $n(\Sigma xy) - (\Sigma x)(\Sigma y) = 10(1158.22) - (286.14)(36) = 1281.16$
   $n(\Sigma x^2) - (\Sigma x)^2 = 10(9260.3968) - (286.14)^2 = 10727.8684$
   $n(\Sigma y^2) - (\Sigma y)^2 = 10(156) - (36)^2 = 264$
   $r = [n(\Sigma xy) - (\Sigma x)(\Sigma y)]/[\sqrt{n(\Sigma x^2) - (\Sigma x)^2} \cdot \sqrt{n(\Sigma y^2) - (\Sigma y)^2}]$
   $= 1281.16/[\sqrt{10727.8684} \cdot \sqrt{264}]$
   $= .761$

c. $H_o: \rho = 0$
   $H_1: \rho \neq 0$
   $\alpha = .05$
   C.R. $r < -.632$ <u>OR</u> C.R. $t < -t_{8,.025} = -2.306$
          $r > .632$ $t > t_{8,.025} = 2.306$
   calculations: calculations:
       $r = .761$ $t_r = (r - \mu_r)/s_r$
                              $= (.761 - 0)/\sqrt{(1-(.761)^2)/8}$
                              $= .761/.229$
                              $= 3.321$

conclusion:
    Reject H_o; there is sufficient evidence to conclude that $\rho \neq 0$ (in fact, $\rho > 0$).

9. a.

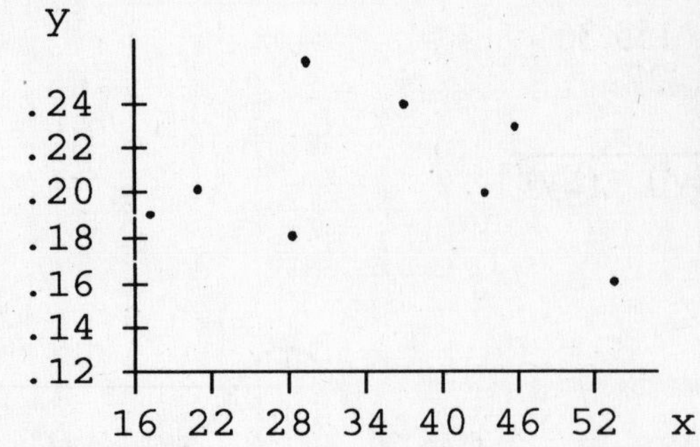

$n = 8$
$\Sigma x = 276.6$
$\Sigma y = 1.66$
$\Sigma xy = 57.191$
$\Sigma x^2 = 10680.48$
$\Sigma y^2 = .3522$

b. $n(\Sigma xy) - (\Sigma x)(\Sigma y) = 8(57.191) - (276.6)(1.66) = -1.628$
$n(\Sigma x^2) - (\Sigma x)^2 = 8(10680.48) - (276.6)^2 = 8936.28$
$n(\Sigma y^2) - (\Sigma y)^2 = 8(.3522) - (1.66)^2 = .062$
$r = [n(\Sigma xy) - (\Sigma x)(\Sigma y)]/[\sqrt{n(\Sigma x^2) - (\Sigma x)^2} \cdot \sqrt{n(\Sigma y^2) - (\Sigma y)^2}]$
$= -1.628/[\sqrt{8936.28} \cdot \sqrt{.062}]$
$= -.069$

c. $H_o: \rho = 0$
$H_1: \rho \neq 0$
$\alpha = .05$
C.R. $r < -.707$ OR C.R. $t < -t_{6,.025} = -2.447$
$r > .707$ $t > t_{6,.025} = 2.447$
calculations: calculations:
$r = -.069$ $t_r = (r - \mu_r)/s_r$
$= (-.069 - 0)/\sqrt{(1-(-.069)^2)/6}$
$= -.069/.407$
$= -.170$

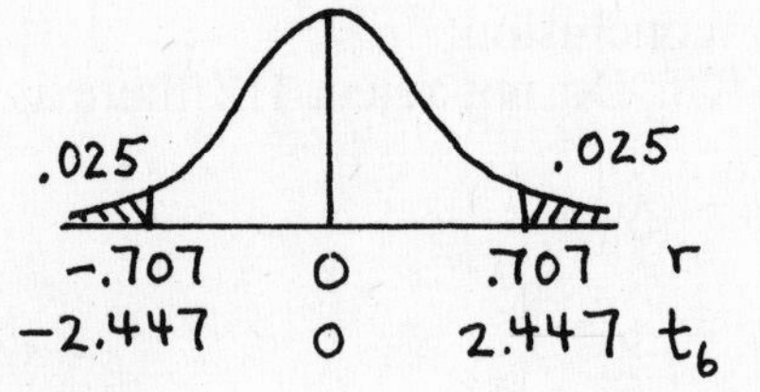

conclusion:
Do not reject H_o; there is not sufficient evidence to conclude that $\rho \neq 0$.

11. a.

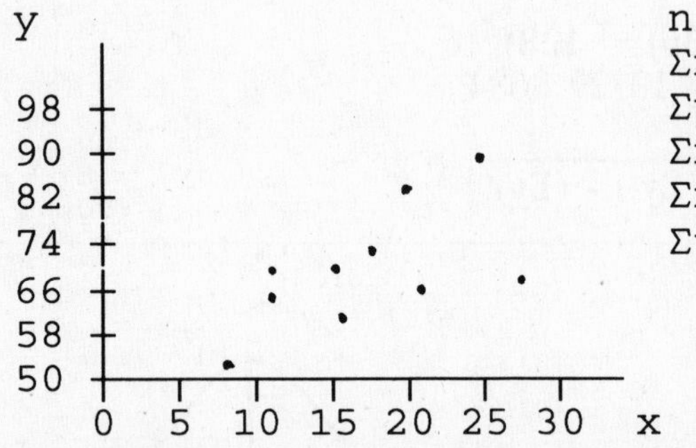

$n = 10$
$\Sigma x = 169$
$\Sigma y = 699$
$\Sigma xy = 12146$
$\Sigma x^2 = 3183$
$\Sigma y^2 = 49927$

b. $n(\Sigma xy) - (\Sigma x)(\Sigma y) = 10(12146) - (169)(699) = 3329$
$n(\Sigma x^2) - (\Sigma x)^2 = 10(3183) - (169)^2 = 3269$
$n(\Sigma y^2) - (\Sigma y)^2 = 10(49927) - (699)^2 = 10669$
$r = [n(\Sigma xy) - (\Sigma x)(\Sigma y)]/[\sqrt{n(\Sigma x^2) - (\Sigma x)^2} \cdot \sqrt{n(\Sigma y^2) - (\Sigma y)^2}]$
$= 3329/[\sqrt{3269} \cdot \sqrt{10669}]$
$= .564$

c. H_o: $\rho = 0$
   H_1: $\rho \neq 0$
   $\alpha = .05$
   C.R. r < -.632   <u>OR</u> C.R. t < $-t_{8,.025}$ = -2.306
        r > .632   t > $t_{8,.025}$ = 2.306
   calculations: calculations:
        r = .564 $t_r = (r - \mu_r)/s_r$
                                      $= (.564 - 0)/\sqrt{(1-(.564)^2)/8}$
                                      $= .564/.292$
                                      $= 1.930$

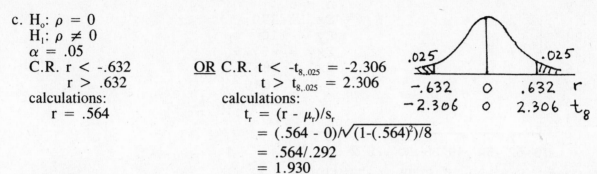

conclusion:
   Do not reject H_o; there is not sufficient evidence to conclude that $\rho \neq 0$.

13. a.

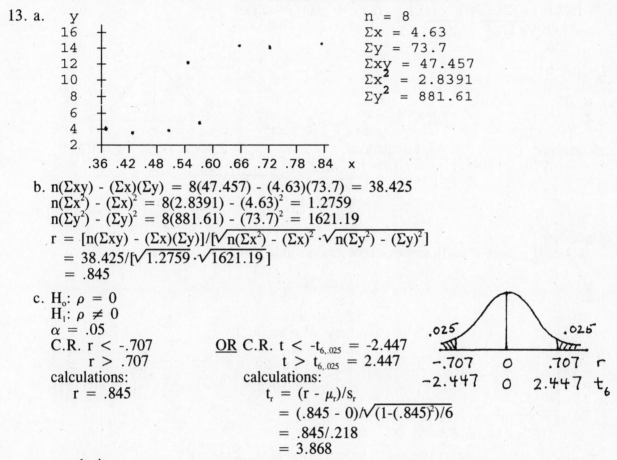

   n = 8
   $\Sigma x = 4.63$
   $\Sigma y = 73.7$
   $\Sigma xy = 47.457$
   $\Sigma x^2 = 2.8391$
   $\Sigma y^2 = 881.61$

b. $n(\Sigma xy) - (\Sigma x)(\Sigma y) = 8(47.457) - (4.63)(73.7) = 38.425$
   $n(\Sigma x^2) - (\Sigma x)^2 = 8(2.8391) - (4.63)^2 = 1.2759$
   $n(\Sigma y^2) - (\Sigma y)^2 = 8(881.61) - (73.7)^2 = 1621.19$
   $r = [n(\Sigma xy) - (\Sigma x)(\Sigma y)]/[\sqrt{n(\Sigma x^2) - (\Sigma x)^2} \cdot \sqrt{n(\Sigma y^2) - (\Sigma y)^2}]$
      $= 38.425/[\sqrt{1.2759} \cdot \sqrt{1621.19}]$
      $= .845$

c. H_o: $\rho = 0$
   H_1: $\rho \neq 0$
   $\alpha = .05$
   C.R. r < -.707   <u>OR</u> C.R. t < $-t_{6,.025}$ = -2.447
        r > .707   t > $t_{6,.025}$ = 2.447
   calculations: calculations:
        r = .845 $t_r = (r - \mu_r)/s_r$
                                      $= (.845 - 0)/\sqrt{(1-(.845)^2)/6}$
                                      $= .845/.218$
                                      $= 3.868$

conclusion:
   Reject H_o; there is sufficient evidence to conclude that $\rho \neq 0$ (in fact, $\rho > 0$).

15. a.

$$n = 8$$
$$\Sigma x = -370$$
$$\Sigma y = 110$$
$$\Sigma xy = -5729$$
$$\Sigma x^2 = 18578$$
$$\Sigma y^2 = 1840$$

b. $n(\Sigma xy) - (\Sigma x)(\Sigma y) = 8(-5729) - (-370)(110) = -5132$

$n(\Sigma x^2) - (\Sigma x)^2 = 8(18578) - (-370)^2 = 11724$

$n(\Sigma y^2) - (\Sigma y)^2 = 8(1840) - (110)^2 = 2620$

$r = [n(\Sigma xy) - (\Sigma x)(\Sigma y)]/[\sqrt{n(\Sigma x^2) - (\Sigma x)^2} \cdot \sqrt{n(\Sigma y^2) - (\Sigma y)^2}]$

$= -5132/[\sqrt{11724} \cdot \sqrt{2620}]$

$= -.926$

c. $H_o: \rho = 0$

$H_1: \rho \neq 0$

$\alpha = .05$

C.R. $r < -.707$ OR C.R. $t < -t_{6,.025} = -2.447$

$r > .707$ $t > t_{6,.025} = 2.447$

calculations: calculations:

$r = -.926$ $t_r = (r - \mu_r)/s_r$

$= (-.926 - 0)/\sqrt{(1-(-.926)^2)/6}$

$= -.926/.154$

$= -6.007$

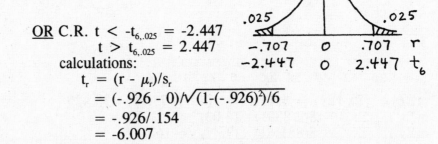

conclusion:

   Reject H_o; there is sufficient evidence to conclude that $\rho \neq 0$ (in fact, $\rho < 0$).

17. a.

$$n = 26$$
$$\Sigma x = 1747.8$$
$$\Sigma y = 916.2$$
$$\Sigma xy = 61354.14$$
$$\Sigma x^2 = 117715.86$$
$$\Sigma y^2 = 56357.08$$

b. $n(\Sigma xy) - (\Sigma x)(\Sigma y) = 26(61354.14) - (1747.8)(916.2) = -6126.72$

$n(\Sigma x^2) - (\Sigma x)^2 = 26(117715.86) - (1747.8)^2 = 5807.52$

$n(\Sigma y^2) - (\Sigma y)^2 = 26(56357.08) - (916.2)^2 = 625861.64$

$r = [n(\Sigma xy) - (\Sigma x)(\Sigma y)]/[\sqrt{n(\Sigma x^2) - (\Sigma x)^2} \cdot \sqrt{n(\Sigma y^2) - (\Sigma y)^2}]$

$= -6126.72/[\sqrt{5807.52} \cdot \sqrt{625861.64}]$

$= -.102$

c. H_o: $\rho = 0$
   H_1: $\rho \neq 0$
   $\alpha = .05$
   C.R. $r < -.396$ OR C.R. $t < -t_{24,.025} = -2.064$
   $\quad\quad r > .396$ $t > t_{24,.025} = 2.064$
   calculations: calculations:
   $\quad r = -.102$ $t_r = (r - \mu_r)/s_r$
                               $= (-.102 - 0)/\sqrt{(1-(-.102)^2)/24}$
                               $= -.102/.203$
                               $= -.500$

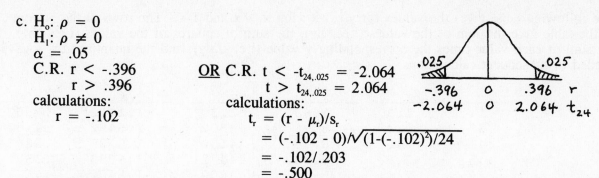

conclusion:
   Do not reject H_o; there is not sufficient evidence to conclude that $\rho \neq 0$.

19. a.

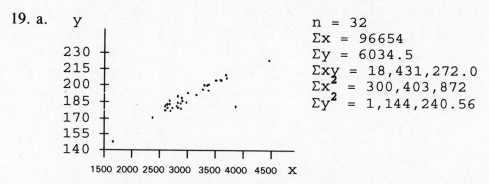

$n = 32$
$\Sigma x = 96654$
$\Sigma y = 6034.5$
$\Sigma xy = 18,431,272.0$
$\Sigma x^2 = 300,403,872$
$\Sigma y^2 = 1,144,240.56$

b. $n(\Sigma xy) - (\Sigma x)(\Sigma y) = 32(18,431,272.0) - (96654)(6034.5) = 6542141.0$
   $n(\Sigma x^2) - (\Sigma x)^2 = 32(300,403,872) - (96654)^2 = 270928188$
   $n(\Sigma y^2) - (\Sigma y)^2 = 32(1,144,240.56) - (6034.5)^2 = 200507.67$
   $r = [n(\Sigma xy) - (\Sigma x)(\Sigma y)]/[\sqrt{n(\Sigma x^2) - (\Sigma x)^2} \cdot \sqrt{n(\Sigma y^2) - (\Sigma y)^2}]$
   $\quad = 6542141.0/[\sqrt{270928188} \cdot \sqrt{200507.67}]$
   $\quad = .888$

c. H_o: $\rho = 0$
   H_1: $\rho \neq 0$
   $\alpha = .05$
   C.R. $r < -.361$ OR C.R. $t < -t_{30,.025} = -1.960$
   $\quad\quad r > .361$ $t > t_{30,.025} = 1.960$
   calculations: calculations:
   $\quad r = .888$ $t_r = (r - \mu_r)/s_r$
                               $= (.888 - 0)/\sqrt{(1-(.888)^2)/30}$
                               $= .888/.0841$
                               $= 10.556$

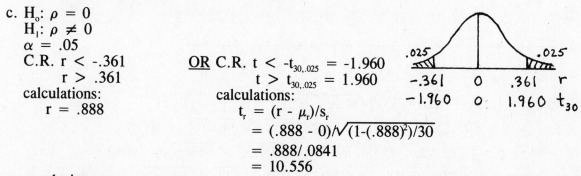

conclusion:
   Reject H_o; there is sufficient evidence to conclude that $\rho \neq 0$ (in fact, $\rho > 0$).

21. A linear correlation coefficient very close to zero indicates <u>no</u> significant linear correlation and no tendencies can be inferred.

23. A linear correlation coefficient very close to zero indicates no significant <u>linear</u> correlation, but there may some other type of relationship between the variables.

25. The following table gives the values for y, x, x^2, log x, $\sqrt{x}$ and 1/x. The rows at the bottom of the table give the sum of the values (i.e., Σv), the sum of squares of the values (i.e., Σv^2), the sum of each value times the corresponding y value (i.e., Σvy), and the quantity $n\Sigma v^2 - (\Sigma v)^2$ needed in subsequent calculations.

y	x	x^2	log x	$\sqrt{x}$	1/x
.11	1.3	1.69	.1139	1.1402	.7692
.38	2.4	5.76	.3802	1.5492	.4167
.41	2.6	6.76	:4150	1.6125	.3846
.45	2.8	7.84	.4472	1.6733	.3571
.39	2.4	5.76	.3802	1.5492	.4167
.48	3.0	9.00	.4771	1.7321	.3333
.61	4.1	16.81	.6128	2.0248	.2439
Σv 2.83	18.6	53.62	2.8264	11.2814	2.9216
Σv^2 1.2817	53.62	539.95	1.2774	18.6000	1.3850
Σvy	8.258	25.495	1.2795	4.7989	1.0326
$n\Sigma v^2 - (\Sigma v)^2$ .9630	29.38	904.55	.9533	2.9300	1.1593

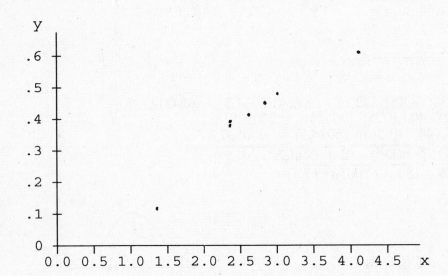

In general, $r = [n(\Sigma vy) - (\Sigma v)(\Sigma y)]/[\sqrt{n(\Sigma v^2) - (\Sigma v)^2} \cdot \sqrt{n(\Sigma y^2) - (\Sigma y)^2}]$

a. For v = x, $r = [7(8.258) - (18.6)(2.83)]/[\sqrt{29.38} \cdot \sqrt{.9630}] = .9716$

b. For v = x^2, $r = [7(25.495) - (53.62)(2.83)]/[\sqrt{904.55} \cdot \sqrt{.9630}] = .9053$

c. For v = log x, $r = [7(1.2795) - (2.8264)(2.83)]/[\sqrt{.9533} \cdot \sqrt{.9630}] = .9996$

d. For v = $\sqrt{x}$, $r = [7(4.7989) - (11.2814)(2.83)]/[\sqrt{2.9300} \cdot \sqrt{.9630}] = .9918$

e. For v = 1/x, $r = [7(1.0326) - (2.9216)(2.83)]/[\sqrt{29.38} \cdot \sqrt{1.1593}] = -.9842$

In each case above, the critical values from Table A-6 for testing significance at the .05 level are $\pm.754$. While all the correlations are significant, the largest value for r was obtained in part (c).

27.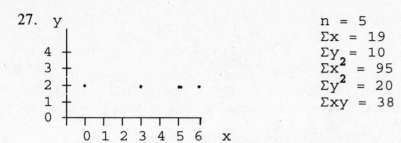

n = 5
Σx = 19
Σy = 10
Σx^2 = 95
Σy^2 = 20
Σxy = 38

$r = [n(\Sigma xy) - (\Sigma x)(\Sigma y)] / [\sqrt{n(\Sigma x^2) - (\Sigma x)^2} \cdot \sqrt{n(\Sigma y^2) - (\Sigma y)^2}]$
  $= [5(38) - (19)(10)] / [\sqrt{5(95) - (19)^2} \cdot \sqrt{5(20) - (10)^2}]$
  $= [0] / [\sqrt{114} \cdot \sqrt{0}]$
  $= 0/0$ which is not defined (officially called an "indeterminate form")

This is an interesting exercise that illustrates what typically happens in mathematics when two opposing lines of reasoning apply.
  * *Since all the points are on the same straight line, the linear correlation is perfect and r must equal 1.00 (or -1.00?).*
  * *Since y is always equal to 2 regardless of x, there is no relationship of any kind between the variables and r must equal 0.*
The result is an indeterminate form, in this case 0/0, and a problem that requires advanced mathematical considerations. While the proper answer is "the linear correlation coefficient does not exist for these data," the generally accepted practical answer is "the linear correlation coefficient for these data may be defined to be zero."
NOTE: The form 0/0 is called "indeterminate" and cannot be permanently assigned a mathematical value because it combines the two opposing truths "0/a = 0" and "a/0 = ∞." Another indeterminate form that cannot be permanently assigned a mathematical value is 0^0, which combines the two opposing truths "$0^a = 0$" and "$a^0 = 1$."

29. a.

n = 8
Σx = -370
Σy = 281
Σxy = -17015
Σx^2 = 18578
Σy^2 = 37579

b. $n(\Sigma xy) - (\Sigma x)(\Sigma y) = 8(-17015) - (-370)(281) = -32150$
  $n(\Sigma x^2) - (\Sigma x)^2 = 8(18578) - (-370)^2 = 11724$
  $n(\Sigma y^2) - (\Sigma y)^2 = 8(37579) - (281)^2 = 221671$
  $r = [n(\Sigma xy) - (\Sigma x)(\Sigma y)] / [\sqrt{n(\Sigma x^2) - (\Sigma x)^2} \cdot \sqrt{n(\Sigma y^2) - (\Sigma y)^2}]$
    $= -32150 / [\sqrt{11724} \cdot \sqrt{221671}]$
    $= -.631$

c. H_o: $\rho = 0$
   H_1: $\rho \neq 0$
   $\alpha = .05$
   C.R. r < -.707   OR C.R. t < $-t_{6,.025}$ = -2.447
         r > .707   t > $t_{6,.025}$ = 2.447
   calculations: calculations:
         r = -.631 $t_r = (r - \mu_r)/s_r$
                        $= (-.631 - 0)/\sqrt{(1-(-.631)^2)/6}$
                        $= -.631/.317$
                        $= -1.991$

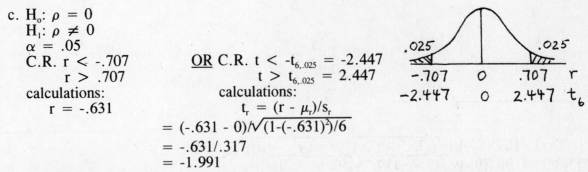

conclusion:
   Do not reject H_o; there is not sufficient evidence to conclude that $\rho \neq 0$.

In this instance, the extreme score lowered to calculated value of r [from -.926] so that it was no longer statistically significant. In general, an extreme score can either lower or raise the calculated value of r -- and the effect can be anywhere from dramatic to minimal.

31.

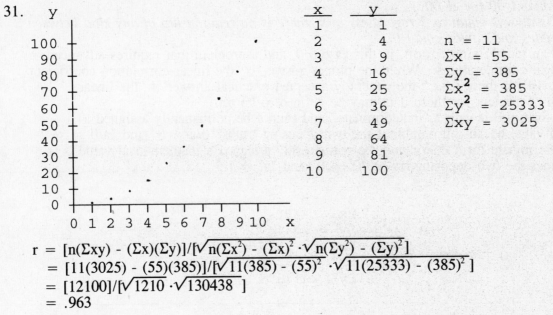

x	y
1	1
2	4
3	9
4	16
5	25
6	36
7	49
8	64
9	81
10	100

n = 11
Σx = 55
Σy = 385
Σx^2 = 385
Σy^2 = 25333
Σxy = 3025

$r = [n(\Sigma xy) - (\Sigma x)(\Sigma y)]/[\sqrt{n(\Sigma x^2) - (\Sigma x)^2} \cdot \sqrt{n(\Sigma y^2) - (\Sigma y)^2}]$
$= [11(3025) - (55)(385)]/[\sqrt{11(385) - (55)^2} \cdot \sqrt{11(25333) - (385)^2}]$
$= [12100]/[\sqrt{1210} \cdot \sqrt{130438}]$
$= .963$

The r = .963 indicates a significant linear relationship -- at the .05 level, the critical values from Table A-6 are $\pm.602$. This section of the parabola, which does not include points on both sides of the minimum at (0,0), is very close to being a straight line.

9-3 Regression

NOTE: For exercises 1-20, the exact summary statistics (i.e., without any rounding) are given on the right. While the intermediate calculations on the left are presented rounded to various degrees of accuracy, the entire unrounded values were preserved in the calculator until the very end.

1. $\bar{x} = 3.50$
   $\bar{y} = 2.17$
   $\hat{\beta}_1 = [n(\Sigma xy) - (\Sigma x)(\Sigma y)]/[n(\Sigma x^2) - (\Sigma x)^2]$
   $= [6(33)-(21)(13)]/[6(85)-(21)^2]$
   $= -75/69$
   $= -1.09$
   $\hat{\beta}_0 = \bar{y} - \hat{\beta}_1\bar{x}$
   $= 2.17 - (-1.09)(3.50)$
   $= 5.97$
   $\hat{y} = \hat{\beta}_0 + \hat{\beta}_1 x$
   $= 5.97 - 1.09x$

   $n = 6$
   $\Sigma x = 21$
   $\Sigma y = 13$
   $\Sigma x^2 = 85$
   $\Sigma y^2 = 43$
   $\Sigma xy = 33$

3. $\bar{x} = 1.75$
   $\bar{y} = 3.00$
   $\hat{\beta}_1 = [n(\Sigma xy) - (\Sigma x)(\Sigma y)]/[n(\Sigma x^2) - (\Sigma x)^2]$
   $= [4(20)-(7)(12)]/[4(15)-(7)^2]$
   $= -4/11$
   $= -.364$
   $\hat{\beta}_0 = \bar{y} - \hat{\beta}_1\bar{x}$
   $= 3.00 - (-.364)(1.75)$
   $= 3.64$
   $\hat{y} = \hat{\beta}_0 + \hat{\beta}_1 x$
   $= 3.64 - .364x$

   $n = 4$
   $\Sigma x = 7$
   $\Sigma y = 12$
   $\Sigma x^2 = 15$
   $\Sigma y^2 = 46$
   $\Sigma xy = 20$

5. $\bar{x} = 7.785$
   $\bar{y} = 3.250$
   $\hat{\beta}_1 = [n(\Sigma xy) - (\Sigma x)(\Sigma y)]/[n(\Sigma x^2) - (\Sigma x)^2]$
   $= [8(221.83)-(62.28)(26)]/[8(533.65)-(62.28)^2]$
   $= 155.36/390.43$
   $= .398$
   $\hat{\beta}_0 = \bar{y} - \hat{\beta}_1\bar{x}$
   $= 3.250 - (.398)(7.785)$
   $= .152$
   $\hat{y} = \hat{\beta}_0 + \hat{\beta}_1 x$
   $= .152 + .398x$

   $n = 8$
   $\Sigma x = 62.28$
   $\Sigma y = 26$
   $\Sigma x^2 = 533.6532$
   $\Sigma y^2 = 104$
   $\Sigma xy = 221.83$

7. $\bar{x} = 28.614$
   $\bar{y} = 3.600$
   $\hat{\beta}_1 = [n(\Sigma xy) - (\Sigma x)(\Sigma y)]/[n(\Sigma x^2) - (\Sigma x)^2]$
   $= [10(1158.22)-(286.14)(36)]/[10(9260.39)-(286.14)^2]$
   $= 1281.16/10727.87$
   $= .119$
   $\hat{\beta}_0 = \bar{y} - \hat{\beta}_1\bar{x}$
   $= 3.600 - (.119)(28.614)$
   $= .183$
   $\hat{y} = \hat{\beta}_0 + \hat{\beta}_1 x$
   $= .183 + .119x$

   $n = 10$
   $\Sigma x = 286.14$
   $\Sigma y = 36$
   $\Sigma x^2 = 9260.3968$
   $\Sigma y^2 = 156$
   $\Sigma xy = 1158.22$

9. $\bar{x}$ = 34.575   n = 8
   $\bar{y}$ = .2075 Σx = 276.6
   $\hat{\beta}_1 = [n(\Sigma xy) - (\Sigma x)(\Sigma y)]/[n(\Sigma x^2) - (\Sigma x)^2]$ Σy = 1.66
   $\quad = [8(57.19)-(276.6)(1.66)]/[8(10680.48)-(276.6)^2]$ Σx^2 = 10680.48
   $\quad = -1.63/8936.28$ Σy^2 = .3522
   $\quad = -.000182$ Σxy = 57.191
   $\hat{\beta}_0 = \bar{y} - \hat{\beta}_1\bar{x}$
   $\quad = .2075 - (-.000182)(34.575)$
   $\quad = .214$
   $\hat{y} = \hat{\beta}_0 + \hat{\beta}_1 x$
   $\quad = .214 - .000182x$

11. $\bar{x}$ = 16.9   n = 10
    $\bar{y}$ = 69.9 Σx = 169
    $\hat{\beta}_1 = [n(\Sigma xy) - (\Sigma x)(\Sigma y)]/[n(\Sigma x^2) - (\Sigma x)^2]$ Σy = 699
    $\quad = [10(12146)-(169)(699)]/[10(3183)-(169)^2]$ Σx^2 = 3183
    $\quad = 3329/3269$ Σy^2 = 49927
    $\quad = 1.02$ Σxy = 12146
    $\hat{\beta}_0 = \bar{y} - \hat{\beta}_1\bar{x}$
    $\quad = 69.9 - (1.02)(16.9)$
    $\quad = 52.7$
    $\hat{y} = \hat{\beta}_0 + \hat{\beta}_1 x$
    $\quad = 52.7 - 1.02x$

13. $\bar{x}$ = .579   n = 8
    $\bar{y}$ = 9.2125 Σx = 4.63
    $\hat{\beta}_1 = [n(\Sigma xy) - (\Sigma x)(\Sigma y)]/[n(\Sigma x^2) - (\Sigma x)^2]$ Σy = 73.7
    $\quad = [8(47.46)-(4.63)(73.7)]/[8(2.84)-(4.63)^2]$ Σx^2 = 2.8391
    $\quad = 38.425/1.276$ Σy^2 = 881.61
    $\quad = 30.12$ Σxy = 47.457
    $\hat{\beta}_0 = \bar{y} - \hat{\beta}_1\bar{x}$
    $\quad = 9.2125 - (30.12)(.579)$
    $\quad = -8.22$
    $\hat{y} = \hat{\beta}_0 + \hat{\beta}_1 x$
    $\quad = -8.22 + 30.12x$
    $\hat{y}_{.75} = -8.22 + 30.12(.75)$
    $\quad = 14.4$

15. $\bar{x}$ = -46.25   n = 8
    $\bar{y}$ = 13.75 Σx = -370
    $\hat{\beta}_1 = [n(\Sigma xy) - (\Sigma x)(\Sigma y)]/[n(\Sigma x^2) - (\Sigma x)^2]$ Σy = 110
    $\quad = [8(-5729)-(-370)(110)]/[8(18578)-(-370)^2]$ Σx^2 = 18578
    $\quad = -5132/11724$ Σy^2 = 1840
    $\quad = -.438$ Σxy = -5729
    $\hat{\beta}_0 = \bar{y} - \hat{\beta}_1\bar{x}$
    $\quad = 13.75 - (-.438)(-46.25)$
    $\quad = -6.50$
    $\hat{y} = \hat{\beta}_0 + \hat{\beta}_1 x$
    $\quad = -6.50 - .438x$
    $\hat{y}_{-60} = -6.50 - .438(-60)$
    $\quad = 19.8$

17. $\bar{x} = 67.22$
$\bar{y} = 35.24$
$\hat{\beta}_1 = [n(\Sigma xy) - (\Sigma x)(\Sigma y)]/[n(\Sigma x^2) - (\Sigma x)^2]$
$= [26(61354)-(1747)(916)]/[26(56357)-(1747)^2]$
$= -6126.72/5807.52$
$= -1.05$
$\hat{\beta}_0 = \bar{y} - \hat{\beta}_1\bar{x}$
$= 35.24 - (-1.05)(67.22)$
$= 106.2$
$\hat{y} = \hat{\beta}_0 + \hat{\beta}_1 x$
$= 106.2 - 1.05x$

$n = 26$
$\Sigma x = 1747.8$
$\Sigma y = 916.2$
$\Sigma x^2 = 117715.86$
$\Sigma y^2 = 56357.08$
$\Sigma xy = 61354.14$

19. $\bar{x} = 188.6$
$\bar{y} = 3020.4$
$\hat{\beta}_1 = [n(\Sigma xy) - (\Sigma x)(\Sigma y)]/[n(\Sigma x^2) - (\Sigma x)^2]$
$= [32(18,431,272)-(6034)(96654)]/[32(1,144,240)...$
$= 6542141/200508 \qquad ...-(6034)^2]$
$= 32.6$
$\hat{\beta}_0 = \bar{y} - \hat{\beta}_1\bar{x}$
$= 3020.4 - (32.6)(188.6)$
$= 3132.5$
$\hat{y} = \hat{\beta}_0 + \hat{\beta}_1 x$
$= 3132.5 + 32.6x$

$n = 32$
$\Sigma x = 6034.5$
$\Sigma y = 96654$
$\Sigma x^2 = 1,144,240.56$
$\Sigma y^2 = 300,403,872$
$\Sigma xy = 18,431,272.0$

21. The .05 critical values for r are taken from Table A-6.
a. CV $= \pm.196$; r$=.999$ is significant
use $\hat{y} = 10.0 + 50x$
$\hat{y}_{2.0} = 10.0 + 50(2.0)$
$= 110.0$

b. CV $= \pm.632$; r$=.005$ is not significant
use $\hat{y} = \bar{y}$
$\hat{y}_{2.0} = \bar{y}$
$= 25.0$

c. CV $= \pm.514$; r$=.519$ is significant
use $\hat{y} = 10.0 + 50x$
$\hat{y}_{2.0} = 10.0 + 50(2.0)$
$= 110.0$

d. CV $= \pm.396$; r$=.393$ is not significant
use $\hat{y} = \bar{y}$
$\hat{y}_{2.0} = \bar{y}$
$= 25.0$

e. CV $= \pm.444$; r$=.567$ is significant
use $\hat{y} = 10.0 + 50x$
$\hat{y}_{2.0} = 10.0 + 50(2.0)$
$= 110.0$

23. The .01 critical values for r are taken from Table A-6.
   a. $CV = \pm.402$; r=.01 is not significant
      use $\hat{y} = \bar{y}$
$$\hat{y}_5 = \bar{y}$$
$$= 6$$

   b. $CV = \pm.402$; r=.93 is significant
      use $\hat{y} = 2 + 3x$
$$\hat{y}_5 = 2 + 3(5)$$
$$= 17$$

   c. $CV = \pm.561$; r=-.654 is significant
      use $\hat{y} = 2 - 3x$
$$\hat{y}_6 = 2 - 3(6)$$
$$= -16$$

   d. $CV = \pm.561$; r=.432 is not significant
      use $\hat{y} = \bar{y}$
$$\hat{y}_5 = \bar{y}$$
$$= 6$$

   e. $CV = \pm.256$; r=-.175 is not significant
      use $\hat{y} = \bar{y}$
$$\hat{y}_5 = \bar{y}$$
$$= 6$$

25. <u>original data</u>

   $n=5$
   $\Sigma x = 4{,}234{,}178$
   $\Sigma y = 576$
   $\Sigma x^2 = 3{,}595{,}324{,}583{,}102$
   $\Sigma y^2 = 67552$
   $\Sigma xy = 491{,}173{,}342$

   $\bar{x} = 846835.6$
   $\bar{y} = 115.2$
   $n\Sigma xy - (\Sigma x)(\Sigma y) = 16{,}980{,}182$
   $n\Sigma x^2 - (\Sigma x)^2 = 48{,}459{,}579{,}826$
   $\hat{\beta}_1 = 16{,}980{,}182/48{,}459{,}579{,}826$
      $= .0003504$
   $\hat{\beta}_0 = \bar{y} - \hat{\beta}_1\bar{x}$
      $= 115.2 - .0003504(846835.6)$
      $= -181.53$
   $\hat{y} = \hat{\beta}_0 + \hat{\beta}_1 x$
      $= -181.53 + .0003504x$

<u>original data divided by 1000</u>

   $n = 5$
   $\Sigma x = 4{,}234.178$
   $\Sigma y = 576$
   $\Sigma x^2 = 3{,}595{,}324.583102$
   $\Sigma y^2 = 67552$
   $\Sigma xy = 491{,}173.342$

   $\bar{x} = 846.8356$
   $\bar{y} = 115.2$
   $n\Sigma xy - (\Sigma x)(\Sigma y) = 16{,}980.182$
   $n\Sigma x^2 - (\Sigma x)^2 = 48{,}459.579826$
   $\hat{\beta}_1 = 16{,}980.182/48{,}459.579826$
      $= .3504$
   $\hat{\beta}_0 = \bar{y} - \hat{\beta}_1\bar{x}$
      $= 115.2 - .3504(846.8356)$
      $= -181.53$
   $\hat{y} = \hat{\beta}_0 + \hat{\beta}_1 x$
      $= -181.53 + .3504x$

Dividing each x by 1000 multiplies $\hat{\beta}_1$, the coefficient of x in the regression equation, by 1000; multiplying the x coefficient by 1000 and dividing x by 1000 will "cancel out" and all predictions remain the same.
Dividing each y by 1000 divides both $\hat{\beta}_1$ and $\hat{\beta}_0$ by 1000; consistent with the new "units" for y, all predictions will also turn out divided by 1000.

27. Replace the values x: 4 5 6 7
    with (x^2): 16 25 36 49
    Use the same format employed for this section's previous exercises substituting (x^2) for x.

$\overline{(x^2)} = 31.5$
$\overline{y} = 10.25$
$\hat{\beta}_1 = [n(\Sigma(x^2)y) - (\Sigma(x^2))(\Sigma y)]/[n(\Sigma(x^2)^2 - (\Sigma(x^2)^2]$
$\quad = [4(1599)-(126)(41)]/[4(4578)-(126)^2]$
$\quad = 1230/2436$
$\quad = .505$
$\hat{\beta}_0 = \overline{y} - \hat{\beta}_1\overline{(x^2)}$
$\quad = 10.25 - (.505)(31.5)$
$\quad = -5.66$
$\hat{y} = \hat{\beta}_0 + \hat{\beta}_1 x^2$
$\quad = -5.66 + .505x^2$

$n = 4$
$\Sigma(x^2) = 126$
$\Sigma y = 41$
$\Sigma(x^2)^2 = 4578$
$\Sigma y^2 = 579$
$\Sigma(x^2)y = 1599$

29. NOTE: Because the formula $\hat{\beta}_0 = \overline{y} - \hat{\beta}_1\overline{x}$ was derived assuming that $\overline{y}$ was the correct predicted value for $\overline{x}$, it cannot be used to prove that fact. One must refer back to the original definition of $\hat{\beta}_0$.

$\hat{y} = \hat{\beta}_0 + \hat{\beta}_1 x$

$\hat{y}_{\overline{x}} = \hat{\beta}_0 + \hat{\beta}_1\overline{x}$

$$= \frac{(\Sigma y)(\Sigma x^2) - (\Sigma x)(\Sigma xy)}{n(\Sigma x^2) - (\Sigma x)^2} + \frac{n(\Sigma xy) - (\Sigma x)(\Sigma y)}{n(\Sigma x^2) - (\Sigma x)^2} \cdot (\Sigma x)/n$$

$$= \frac{(\Sigma y)(\Sigma x^2) - (\Sigma x)(\Sigma xy) + [n(\Sigma xy) - (\Sigma x)(\Sigma y)] \cdot (\Sigma x)/n}{n(\Sigma x^2) - (\Sigma x)^2}$$

$$= \frac{(\Sigma y)(\Sigma x^2) - (\Sigma x)(\Sigma xy) + n(\Sigma xy)(\Sigma x)/n - (\Sigma x)(\Sigma y)(\Sigma x)/n}{n(\Sigma x^2) - (\Sigma x)^2}$$

$$= \frac{(\Sigma y)(\Sigma x^2) - (\Sigma x)(\Sigma xy) + (\Sigma xy)(\Sigma x) - (\Sigma y)(\Sigma x)^2/n}{n(\Sigma x^2) - (\Sigma x)^2}$$

$$= \frac{(\Sigma y)(\Sigma x^2) - (\Sigma y)(\Sigma x)^2/n}{n(\Sigma x^2) - (\Sigma x)^2}$$

$$= \frac{(\Sigma y)[(\Sigma x^2) - (\Sigma x)^2/n]}{n[(\Sigma x^2) - (\Sigma x)^2/n]}$$

$$= (\Sigma y)/n$$

$$= \overline{y}$$

9-4 Variation and Prediction Intervals

1. The coefficient of determination is $r^2 = (.333)^2 = .111$.
   The portion of the total variation explained by the regression line is $r^2 = .111 = 11.1\%$.

3. The coefficient of determination is $r^2 = (.800)^2 = .640$.
   The portion of the total variation explained by the regression line is $r^2 = .640 = 64.0\%$.

5. The predicted values were calculated using the regression line $\hat{y} = 2 + 3x$.

x	y	$\hat{y}$	$\bar{y}$	$\hat{y}-\bar{y}$	$(\hat{y}-\bar{y})^2$	$y-\hat{y}$	$(y-\hat{y})^2$	$y-\bar{y}$	$(y-\bar{y})^2$
1	5	5	12.2	-7.2	51.84	0	0	-7.2	51.84
2	8	8	12.2	-4.2	17.64	0	0	-4.2	17.64
3	11	11	12.2	-1.2	1.44	0	0	-1.2	1.44
5	17	17	12.2	4.8	23.04	0	0	4.8	23.04
6	20	20	12.2	7.8	60.84	0	0	7.8	60.84
17	61	61	61.0	0	154.80	0	0	0	154.80

NOTE: A table such as the one above organizes the work and provides all the values needed to discuss variation. In such a table, the following must always be true and can be used as a check before proceeding.
* $\Sigma y = \Sigma \hat{y} = \Sigma \bar{y}$
* $\Sigma(\hat{y}-\bar{y}) = \Sigma(y-\hat{y}) = \Sigma(y-\bar{y}) = 0$
* $\Sigma(y-\bar{y})^2 + \Sigma(y-\hat{y})^2 = \Sigma(y-\bar{y})^2$

a. The explained variation is $\Sigma(\hat{y}-\bar{y})^2 = 154.80$

b. The unexplained variation is $\Sigma(y-\hat{y})^2 = 0$

c. The total variation is $\Sigma(y-\bar{y})^2 = 154.80$

d. $r^2 = \Sigma(\hat{y}-\bar{y})^2/\Sigma(y-\bar{y})^2$
   $= 154.80/154.80$
   $= 1.00$

e. $s_e^2 = \Sigma(y-\hat{y})^2/(n-2)$
   $= 0/3$
   $= 0$
   $s_e = 0$

7. The predicted values were calculated using the regression line $\hat{y} = -8.21713 + 30.1160x$.

x	y	$\hat{y}$	$\overline{y}$	$\hat{y}-\overline{y}$	$(\hat{y}-\overline{y})^2$	$y-\hat{y}$	$(y-\hat{y})^2$	$y-\overline{y}$	$(y-\overline{y})^2$
.65	14.7	11.358	9.2125	2.146	4.604	3.342	11.167	5.4875	30.113
.55	12.3	8.347	9.2125	-.866	.750	3.953	15.629	3.0875	9.533
.72	14.6	13.466	9.2125	4.254	18.096	1.134	1.285	5.3875	29.025
.83	15.1	16.779	9.2125	7.567	57.254	-1.679	2.820	5.8875	34.663
.57	5.0	8.949	9.2125	-.264	.069	-3.949	15.595	-4.2125	17.745
.51	4.1	7.142	9.2125	-2.070	4.287	-3.042	9.254	-5.1125	26.138
.43	3.8	4.733	9.2125	-4.480	20.068	-.933	.870	-5.4125	29.295
.37	4.1	2.926	9.2125	-6.287	39.523	1.174	1.379	-5.1125	26.138
4.63	73.7	73.700	73.0000	0	144.651	0	57.998	0	202.649

a. The explained variation is $\Sigma(\hat{y}-\overline{y})^2 = 144.651$

b. The unexplained variation is $\Sigma(y-\hat{y})^2 = 57.998$

c. The total variation is $\Sigma(y-\overline{y})^2 = 202.649$

d. $r^2 = \Sigma(\hat{y}-\overline{y})^2/\Sigma(y-\overline{y})^2$
$= 144.651/202.649$
$= .714$

e. $s_e^2 = \Sigma(y-\hat{y})^2/(n-2)$
$= 57.998/6$
$= 9.666$
$s_e = 3.109$

9. a. $\hat{y} = 2 + 3x$
$\hat{y}_4 = 2 + 3(4)$
$= 14$

b. $\hat{y} \pm t_{n-2,\alpha/2}s_e\sqrt{1 + 1/n + n(x_o-\overline{x})^2/[n\Sigma x^2-(\Sigma x)^2]}$
$\hat{y} \pm 0$, since $s_e = 0$
The prediction "interval" in this case shrinks to a single point. Since $r^2 = 1.00$ (i.e., 100% of the variability in the y's can be explained by the regression), a perfect prediction can be made. For a practical example of such a situation, consider the regression line $\hat{y} = 1.099x$ for predicting the amount of money y due for purchasing x gallons of gasoline at \$1.099 per gallon -- the "prediction" will be exactly correct every time because there is a perfect correlation between the number of gallons purchased and the amount of money due.

11. a. $\hat{y} = -89.21713 + 30.1160x$
$\hat{y}_{.75} = -8.21713 + 30.1160(.75)$
$= 14.37$

b. preliminary calculations
$n = 8$
$\Sigma x = 4.63$ $\overline{x} = (\Sigma x)/n = 4.63/8 = .57875$
$\Sigma x^2 = 2.8391$ $n\Sigma x^2-(\Sigma x)^2 = 8(2.8391)-(4.63)^2 = 1.2759$
$\hat{y} \pm t_{n-2,\alpha/2}s_e\sqrt{1 + 1/n + n(x_o-\overline{x})^2/[n\Sigma x^2-(\Sigma x)^2]}$
$\hat{y}_{.75} \pm t_{6,.025}(3.109)\sqrt{1 + 1/8 + 8(.75-.57875)^2/[1.2759]}$
$14.37 \pm (2.447)(3.109)\sqrt{1.30888}$
14.37 ± 8.70
$5.67 < y_{.75} < 23.07$

Exercises 13-16 refer to the chapter problem of Table 9-1. They use the following, which are calculated and/or discussed at various places in the text,

$n = 8$

$\Sigma x = 14.60$ $\hat{y} = .549 + 1.480x$

$\Sigma x^2 = 32.9632$ $s_e = .971554$

and the values obtained below.

$\bar{x} = (\Sigma x)/n = 14.60/8 = 1.825$

$n\Sigma x^2 - (\Sigma x)^2 = 8(32.9632) - (14.60)^2 = 50.5456$

13. $\hat{y} \pm t_{n-2,\alpha/2} s_e \sqrt{1 + 1/n + n(x_o - \bar{x})^2/[n\Sigma x^2 - (\Sigma x)^2]}$

   $\hat{y}_{1.00} \pm t_{6,.025}(.971554)\sqrt{1 + 1/8 + 8(1.00 - 1.825)^2/[50.5456]}$

   $2.029 \pm (2.447)(.971554)\sqrt{1.23272}$

   2.029 ± 2.640

   $-.61 < y_{1.00} < 4.67$

15. $\hat{y} \pm t_{n-2,\alpha/2} s_e \sqrt{1 + 1/n + n(x_o - \bar{x})^2/[n\Sigma x^2 - (\Sigma x)^2]}$

   $\hat{y}_{1.50} \pm t_{6,.05}(.971554)\sqrt{1 + 1/8 + 8(1.50 - 1.825)^2/[50.5456]}$

   $2.769 \pm (1.943)(.971554)\sqrt{1.14172}$

   2.769 ± 2.017

   $.75 < y_{1.50} < 4.79$

17. This exercise uses the following values from the chapter problem of Table 9-1, which are calculated and/or discussed at various places in the text,

   $n = 8$ $\hat{\beta}_o = .549$

   $\Sigma x = 14.60$ $\hat{\beta}_1 = 1.480$

   $\Sigma x^2 = 32.9632$ $s_e = .971554$

   and the values obtained below.

   $\bar{x} = (\Sigma x)/n = 14.60/8 = 1.825$

   $\Sigma x^2 - (\Sigma x)^2/n = (32.9632) - (14.60)^2/8 = 6.3182$

   a. $\hat{\beta}_o \pm t_{n-2,\alpha/2} s_e \sqrt{1/n + \bar{x}^2/[\Sigma x^2 - (\Sigma x)^2/n]}$

      $.549 \pm t_{6,.025}(.971554)\sqrt{1/8 + (1.825)^2/[6.3182]}$

      $.549 \pm (2.447)(.971554)\sqrt{.65215}$

      $.549 \pm 1.920$

      $-1.37 < \beta_o < 2.47$

   b. $\hat{\beta}_1 \pm t_{n-2,\alpha/2} s_e/\sqrt{\Sigma x^2 - (\Sigma x)^2/n}$

      $1.480 \pm t_{6,.025}(.971554)/\sqrt{6.3182}$

      $1.480 \pm (2.447)(.971554)/\sqrt{6.3182}$

      $1.480 \pm .946$

      $.53 < \beta_1 < 2.43$

19. a. $s_e^2 = \Sigma(y-\hat{y})^2/(n-2)$

$(n-2)s_e^2 = \Sigma(y-\hat{y})^2$

The unexplained variation is equal to $(n-2)s_e^2$.

b.
$$r^2 = \text{(explained variation)/(total variation)}$$
$$\text{(total variation)} \cdot r^2 = \text{(explained variation)}$$
$$[\text{(explained variation)} + \text{(unexplained variation)}] \cdot r^2 = \text{(explained variation)}$$
$$\text{(explained variation)} \cdot r^2 + \text{(unexplained variation)} \cdot r^2 = \text{(explained variation)}$$
$$\text{(unexplained variation)} \cdot r^2 = \text{(explained variation)} - \text{(explained variation)} \cdot r^2$$
$$\text{(unexplained variation)} \cdot r^2 = \text{(explained variation)}(1 - r^2)$$
$$\text{(unexplained variation)} \cdot r^2/(1 - r^2) = \text{(explained variation)}$$

The explained variation is equal to (unexplained variation) $\cdot r^2/(1 - r^2)$.

c. If $r^2 = .900$, then $r = \pm.949$.

Since the regression line has a negative slope (i.e., $\hat{\beta}_1 = -2$), we choose the negative root. The linear correlation coefficient, therefore, is $r = -.949$.

9-5 Multiple Regression

1. $\hat{y} = 34.8 + 1.21x_1 + .23x_2$

$\hat{y}_{24,92} = 34.8 + 1.21(24) + .23(92)$

$= 85.0$

3. $\hat{y} = 34.8 + 1.21x_1 + .23x_2$

$\hat{y}_{18,81} = 34.8 + 1.21(18) + .23(81)$

$= 75.2$

5. Let y be the household size.

Let x_1, x_2 and x_3 be the weights of discarded metal, plastic and food.

$\hat{y} = .92 - .244x_1 + 1.75x_2 - .073x_3$

7. No; since the P-value of .116 is greater than .05, there is not statistical significance. The equation $\hat{y} = \bar{y}$ should be used for making predictions.

9. Following is the complete Minitab input and output required for the problem.
   NOTE: It is often easier to input the data as whole numbers and then to use a LET statement to move the decimal point accordingly. In addition, it is a good idea to PRINT the data before analyzing it -- just to make certain that Minitab is talking about the same data that you are.

a. $\hat{y} = .804 - .363x_2 + 1.76x_4$

b. P-value = .040

c. $R^2 = 72.3\% = .723$

d. adjusted $R^2 = 61.3\% = .613$

e. yes; since .040 < .05

```
MTB > SET C1
DATA> 2 3 3 6 4 2 1 5
DATA> END
MTB > SET C2
DATA> 109 104 257 302 150 210 193 357
DATA> END
MTB > LET C2=C2/100
MTB > SET C4
DATA> 27 141 219 283 219 181 85 305
DATA> END
MTB > LET C4=C4/100
MTB > NAME C1'Y' C2'X2' C4'X4'
MTB > PRINT C1 C2 C4

   ROW     Y      X2      X4

    1      2     1.09    0.27
    2      3     1.04    1.41
    3      3     2.57    2.19
    4      6     3.02    2.83
    5      4     1.50    2.19
    6      2     2.10    1.81
    7      1     1.93    0.85
    8      5     3.57    3.05

MTB > REGRESS C1 2 C2 C4


The regression equation is
Y = 0.804 - 0.363 X2 + 1.76 X4

Predictor        Coef       Stdev    t-ratio         p
Constant       0.8038      0.9820       0.82     0.450
X2            -0.3635      0.7288      -0.50     0.639
X4             1.7591      0.6959       2.53     0.053

s = 1.039      R-sq = 72.3%      R-sq(adj) = 61.3%

Analysis of Variance

SOURCE        DF          SS         MS         F         p
Regression     2      14.105      7.052      6.54     0.040
Error          5       5.395      1.079
Total          7      19.500

SOURCE        DF      SEQ SS
X2             1       7.211
X4             1       6.894
```

11. Following is the complete Minitab input and output required for the problem.
    NOTE: When a command is given after DATA>, Minitab assumes the ENDOFDATA and
    executes the command. It is a good idea to PRINT the data before analyzing it -- just to
    make certain that Minitab is talking about the same data that you are.

The regression equation is $\hat{y} = 2.17 + 2.44x + .464x^2$.
The multiple coefficient of determination $R^2 = 100.0\%$ means that all the points lie exactly
    (i.e., within the accuracy of the problem) on the parabola.

```
MTB > SET C1
DATA> 5 14 19 42 26
DATA> SET C2
DATA> 1 3 4 7 5
DATA> LET C3=C2*C2
MTB > NAME C1'Y' C2'X' C3'X2'
MTB > PRINT C1-C3

 ROW      Y      X      X2

   1      5      1      1
   2     14      3      9
   3     19      4     16
   4     42      7     49
   5     26      5     25


MTB > REGRESS C1 2 C2 C3

The regression equation is
Y = 2.17 + 2.44 X + 0.464 X2


Predictor         Coef        Stdev      t-ratio          p
Constant        2.1714       0.5741         3.78      0.063
X               2.4357       0.3140         7.76      0.016
X2             0.46429      0.03802        12.21      0.007


s = 0.3485      R-sq = 100.0%    R-sq(adj) = 99.9%

Analysis of Variance

SOURCE          DF           SS          MS         F          p
Regression       2       774.56      387.28   3189.35      0.000
Error            2         0.24        0.12
Total            4       774.80

SOURCE          DF       SEQ SS
X                1       756.45
X2               1        18.11
```

Review Exercises

1. Let x be the cost and y be the miles.

n = 8

$\Sigma x = 190.3$

$\Sigma y = 281.9$

$\Sigma x^2 = 5332.07$

$\Sigma y^2 = 11616.25$

$\Sigma xy = 7073.09$

$n(\Sigma xy) - (\Sigma x)(\Sigma y) = 8(7073.09) - (190.3)(281.9)$
$= 2939.15$

$n(\Sigma x^2) - (\Sigma x)^2 = 8(5332.07) - (190.3)^2$
$= 6642.47$

$n(\Sigma y^2) - (\Sigma y)^2 = 8(11616.25) - (281.9)^2$
$= 13462.39$

a. $r = [n(\Sigma xy) - (\Sigma x)(\Sigma y)]/[\sqrt{n(\Sigma x^2) - (\Sigma x)^2} \cdot \sqrt{n(\Sigma y^2) - (\Sigma y)^2}]$
$= [2939.15]/[\sqrt{6442.47} \cdot \sqrt{13462.39}]$
$= .316$

b. $CV = \pm.707$

c. no significant linear correlation

d. $\hat{\beta}_1 = [n(\Sigma xy) - (\Sigma x)(\Sigma y)]/[n(\Sigma x^2) - (\Sigma x)^2]$
$= 2939.15/6442.47$
$= .456$

$\hat{\beta}_0 = \bar{y} - \hat{\beta}_1\bar{x}$
$= (281.9/8) - (.456)(190.3/8)$
$= 24.4$

$\hat{y} = \hat{\beta}_0 + \hat{\beta}_1 x$
$= 24.4 + .456x$

e.

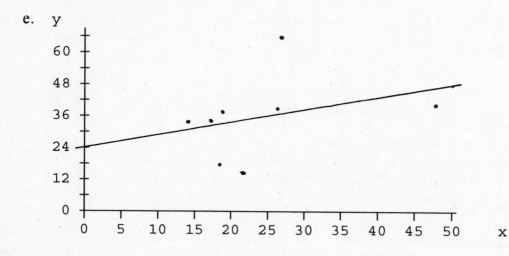

3. Let x be the minutes and y be the points.

n = 9 $\qquad$ $n(\Sigma xy) - (\Sigma x)(\Sigma y) = 9(4,581,424) - (7238)(3741)$

$\Sigma x = 7238$ $\qquad\qquad\qquad = 14,155,458$

$\Sigma y = 3741$ $\qquad$ $n(\Sigma x^2) - (\Sigma x)^2 = 9(8,267,024) - (7238)^2$

$\Sigma x^2 = 8,267,024$ $\qquad\qquad\qquad = 22,014,572$

$\Sigma y^2 = 2,637,573$ $\qquad$ $n(\Sigma y^2) - (\Sigma y)^2 = 9(2,637,573) - (3741)^2$

$\Sigma xy = 4,581,424$ $\qquad\qquad\qquad = 9,743,076$

a. $r = [n(\Sigma xy) - (\Sigma x)(\Sigma y)]/[\sqrt{n(\Sigma x^2) - (\Sigma x)^2} \cdot \sqrt{n(\Sigma y^2) - (\Sigma y)^2}]$

$= [14,155,458]/[\sqrt{22,014,572} \ \sqrt{9,743,076}]$

$= .967$

b. $CV = \pm.666$

c. significant (positive) linear correlation

d. $\hat{\beta}_1 = [n(\Sigma xy) - (\Sigma x)(\Sigma y)]/[n(\Sigma x^2) - (\Sigma x)^2]$

$= 14,155,458/22,014,572$

$= .643$

$\hat{\beta}_0 = \bar{y} - \hat{\beta}_1\bar{x}$

$= (3741/9) - (.643)(7238/9)$

$= -101$

$\hat{y} = \hat{\beta}_0 + \hat{\beta}_1 x$

$= -101 + .643x$

e.

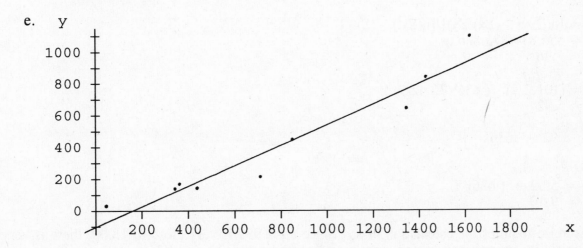

For exercises 5 and 7, let x be the minutes and y be the savings.

$$n = 9 \qquad\qquad n(\Sigma xy) - (\Sigma x)(\Sigma y) = 9(909.927) - (75.42)(101.2)$$
$$\Sigma x = 75.42 \qquad\qquad\qquad\qquad 556.839$$
$$\Sigma y = 101.2 \qquad\qquad n(\Sigma x^2) - (\Sigma x)^2 = 9(729.3314) - (75.42)^2$$
$$\Sigma x^2 = 729.3314 \qquad\qquad\qquad = 875.8062$$
$$\Sigma y^2 = 1177.36 \qquad\qquad n(\Sigma y^2) - (\Sigma y)^2 = 9(1177.36) - (101.2)^2$$
$$\Sigma xy = 909.927 \qquad\qquad\qquad = 354.8$$

5. a. $r = [n(\Sigma xy) - (\Sigma x)(\Sigma y)]/[\sqrt{n(\Sigma x^2) - (\Sigma x)^2} \cdot \sqrt{n(\Sigma y^2) - (\Sigma y)^2}]$
   $= [556.839]/[\sqrt{875.8062} \cdot \sqrt{354.8}\,]$
   $= .999$

   H_o: $\rho = 0$
   H_1: $\rho \neq 0$
   $\alpha = .05$
   C.R. $r < -.666$ OR C.R. $t < -t_{7,.025} = -2.365$
        $r > .666$ $t > t_{7,.025} = 2.365$
   calculations: calculations:
       $r = .999$ $t_r = (r - \mu_r)/s_r$
                      $= (.999 - 0)/\sqrt{(1-(.999)^2)/7}$
                      $= .999/.0175$
                      $= 57.072$

   conclusion:
      Reject H_o; there is sufficient evidence to conclude that $\rho \neq 0$ (in fact, $\rho > 0$).

   b. $\hat{\beta}_1 = [n(\Sigma xy) - (\Sigma x)(\Sigma y)]/[n(\Sigma x^2) - (\Sigma x)^2]$
       $= 556.839/875.8062$
       $= .636$
      $\hat{\beta}_o = \bar{y} - \hat{\beta}_1\bar{x}$
        $= (101.2/9) - (.636)(75.42/9)$
        $= 5.92$
      $\hat{y} = \hat{\beta}_o + \hat{\beta}_1 x$
        $= 5.92 + .636x$

   c. $\hat{y} = \hat{\beta}_o + \hat{\beta}_1 x$
      $\hat{y}_{5.0} = 5.92 + (.636)(5)$
          $= 9.10$

7. This exercise requires the calculation of s_e. Since $r = .999$ is very close to 1.000, there is very little variability unexplained by the regression line and s_e will be very close to zero. This necessitates extra care in the calculations for s_e -- STORE and RECALL with complete accuracy any intermediate values other than n or the primary summations given exactly above.

$$s_e^2 = [\Sigma y^2 - \hat{\beta}_o(\Sigma y) - \hat{\beta}_1(\Sigma xy)]/(n-2)$$
$$= [1177.36 - (5.9164)(101.2) - (.6358)(909.927)]/7$$
$$= .084540/7$$
$$= .012077$$
$$s_e = .1099$$

$$\hat{y} \pm t_{n-2,\alpha/2}s_e\sqrt{1 + 1/n + n(x_o-\bar{x})^2/[n\Sigma x^2-(\Sigma x)^2]}$$
$$\hat{y}_{5.0} \pm t_{7,.025}(.1099)\sqrt{1 + 1/9 + 9(5.0-8.38)^2/[875.8062]}$$
$$9.095 \pm (2.365)(.1099)\sqrt{1.2285}$$
$$9.095 \pm .288$$
$$8.81 < y_{5.0} < 9.38$$

Chapter 10

Multinomial Experiments and Contingency Tables

10-2 Multinomial Experiments

NOTE: In multinomial problems, always verify that $\Sigma E = \Sigma O$ before proceeding. If these sums are not equal, then an error has been made and further calculations have no meaning.

1. H_o: $p_0 = p_1 = p_2 = \ldots = p_9 = .10$
H_1: at least one of the proportions is different from .10
$\alpha = .01$
C.R. $\chi^2 > \chi^2_{9,.01} = 21.666$
calculations:

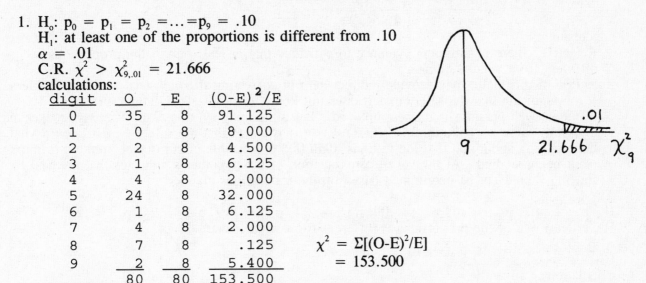

digit	O	E	$(O-E)^2/E$
0	35	8	91.125
1	0	8	8.000
2	2	8	4.500
3	1	8	6.125
4	4	8	2.000
5	24	8	32.000
6	1	8	6.125
7	4	8	2.000
8	7	8	.125
9	2	8	5.400
	80	80	153.500

$\chi^2 = \Sigma[(O-E)^2/E]$
$= 153.500$

conclusion:
Reject H_o; there is sufficient evidence to conclude that at least one of the proportions is different from .10.

3. H_o: $p_{Sun} = p_{Mon} = p_{Tue} = \ldots = p_{Sat} = 1/7$
H_1: at least one of the proportions is different from 1/7
$\alpha = .05$
C.R. $\chi^2 > \chi^2_{6,.05} = 12.592$
calculations:

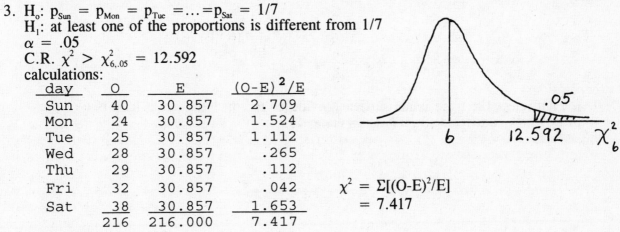

day	O	E	$(O-E)^2/E$
Sun	40	30.857	2.709
Mon	24	30.857	1.524
Tue	25	30.857	1.112
Wed	28	30.857	.265
Thu	29	30.857	.112
Fri	32	30.857	.042
Sat	38	30.857	1.653
	216	216.000	7.417

$\chi^2 = \Sigma[(O-E)^2/E]$
$= 7.417$

conclusion:
Do not reject H_o; there is not sufficient evidence to conclude that at least one of the proportions is different from 1/7.

5. H_o: $p_{Mon} = .30$, $p_{Tue} = .15$, $p_{Wed} = .15$, $p_{Thu} = .20$, $p_{Fri} = .20$
   H_1: at least one of the proportions is different from what is claimed
   $\alpha = .05$
   C.R. $\chi^2 > \chi^2_{4,.05} = 9.488$
   calculations:

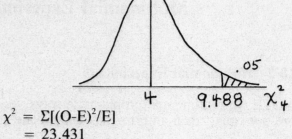

day	O	E	$(O-E)^2/E$
Mon	31	44.10	3.891
Tue	42	22.05	18.050
Wed	18	22.05	.744
Thu	25	29.40	.659
Fri	31	29.40	.087
	147	147.00	23.431

$\chi^2 = \Sigma[(O-E)^2/E]$
$= 23.431$

conclusion:
Reject H_o; there is sufficient evidence to conclude that at least one of the proportions is different from what is claimed.
Rejection of this claim may provide indirect help in correcting the industrial accident problem. It may indicate that the safety expert does not know what he talking about and that a different such person should be employed. If a single day (in this case, Tuesday) makes an unusually large contribution to $\Sigma[(O-E)^2/E]$, it may mean that the safety expert knows what he is talking about and that the test has identified correctable circumstances unique to the plant being studied. At the very least, rejection of the hypothesis indicates that medical staffing should not be determined based on the proportions claimed.

7. H_o: $p_{Sun} = .07$, $p_{Mon} = .05$ $p_{Tue} = .09$, $p_{Wed} = .11$, $p_{Thu} = .19$, $p_{Fri} = .24$, $p_{Sat} = .25$
   H_1: at least one of the proportions is different from what is claimed
   $\alpha = .05$
   C.R. $\chi^2 > \chi^2_{6,.05} = 12.592$
   calculations:

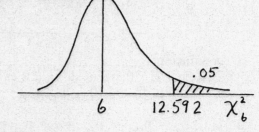

day	O	E	$(O-E)^2/E$
Sun	9	7.21	.444
Mon	6	5.15	.140
Tue	10	9.27	.057
Wed	8	11.33	.979
Thu	19	9.57	.017
Fri	23	24.72	.120
Sat	28	25.75	.197
	103	103.00	1.954

$\chi^2 = \Sigma[(O-E)^2/E]$
$= 1.954$

conclusion:
Do not reject H_o; there is not sufficient evidence to conclude that at least one of the proportions is different from what is claimed.

9. H_o: $p_0 = p_1 = p_2 = ... = p_9 = .10$
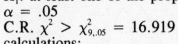
H_1: at least one of the proportions is different from .10
$\alpha = .05$
C.R. $\chi^2 > \chi^2_{9,.05} = 16.919$
calculations:

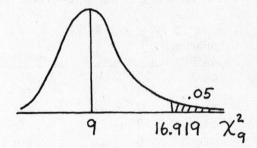

digit	O	E	$(O-E)^2/E$
0	8	10	.400
1	8	10	.400
2	12	10	.400
3	11	10	.100
4	10	10	.000
5	8	10	.400
6	9	10	.100
7	8	10	.400
8	12	10	.400
9	14	10	1.600
	100	100	4.200

$\chi^2 = \Sigma[(O-E)^2/E]$
$= 4.200$

conclusion:
   Do not reject H_o; there is not sufficient evidence to conclude that at least one of the proportions is different from .10.

11. NOTE: Usually the hypothesized p_i's are given and the formula $E_i = np_i$ is used to find the individual expected values. Here, the individual expected values are given (and they sum to $n = 320$) and the formula must be used "in reverse" to solve for the hypothesized p_i's.
   H_o: $p_A = .0625$, $p_B = .0625$, $p_C = .1250$, $p_D = .3750$, $p_E = .3750$
   H_1: at least one of the proportions is different from what is claimed
   $\alpha = .01$
   C.R. $\chi^2 > \chi^2_{4,.01} = 13.277$
   calculations:

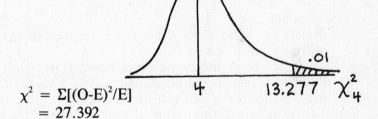

type	O	E	$(O-E)^2/E$
A	30	20	5.000
B	15	20	1.250
C	58	40	8.100
D	83	120	11.408
E	134	120	1.633
	320	320	27.392

$\chi^2 = \Sigma[(O-E)^2/E]$
$= 27.392$

conclusion:
   Reject H_o; there is sufficient evidence to conclude that at least one of the proportions is different from what is claimed.

13. H_o: $p_{bro} = .30$, $p_{yel} = .20$, $p_{red} = .20$, $p_{ora} = .10$, $p_{gre} = .10$, $p_{tan} = .10$
H_1: at least one of the proportions is different from what is claimed
$\alpha = .05$
C.R. $\chi^2 > \chi^2_{5,.05} = 11.071$
calculations:

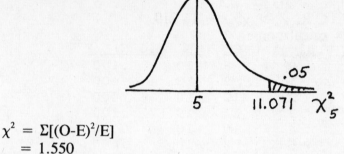

color	O	E	$(O-E)^2/E$
bro	30	30	.000
yel	24	20	.800
red	17	20	.450
ora	9	10	.100
gre	9	10	.100
tan	11	10	.100
	100	100	1.550

$\chi^2 = \Sigma[(O-E)^2/E]$
$= 1.550$

conclusion:
Do not reject H_o; there is not sufficient evidence to conclude that at least one of the proportions is different from what is claimed.

15. H_o: $p_{poor} = p_{fair} = p_{good} = p_{exce} = .25$
H_1: at least one of the proportions is different from .25
$\alpha = .05$
C.R. $\chi^2 > \chi^2_{3,.05} = 7.815$
calculations:

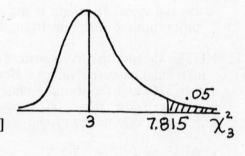

type	O	E	$(O-E)^2/E$
poor	4	15	8.067
fair	13	15	.267
good	29	15	13.067
exce	14	15	.067
	60	60	21.467

$\chi^2 = \Sigma[(O-E)^2/E]$
$= 21.467$

conclusion:
Reject H_o; there is sufficient evidence to conclude that at least one of the proportions is different from .25.

NOTE: Before giving movie critic James Harrington two thumbs down on the basis of the above test, one should determine exactly how the list of movies in the appendix was obtained and whether it is a fair test of the critic's claim. Is it possible, for instance, that the given list was obtained from a list of movies being shown at some point in time? Since poor movies have shorter runs, they would be under represented in the sample. Or is it possible that the critic as talking of "all films produced" and not merely those that make it onto some list. As a recognized critic with inside connections, he might be aware of films so poor that they never make it to the box office or onto lists that are compiled.

16. H_o: $p_0 = p_1 = p_2 = \ldots = p_9 = .10$
H_1: at least one of the proportions is different from .10
$\alpha = .05$
C.R. $\chi^2 > \chi^2_{9,.05} = 16.919$
calculations:

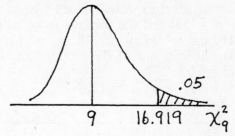

digit	O	E	$(O-E)^2/E$
0	13	15	.267
1	16	15	.067
2	11	15	1.067
3	17	15	.267
4	14	15	.067
5	18	15	.600
6	18	15	.600
7	18	15	.600
8	12	15	.600
9	13	15	.267
	150	150	4.400

$\chi^2 = \Sigma[(O-E)^2/E]$
$= 4.400$

conclusion:
Do not reject H_o; there is not sufficient evidence to conclude that at least one of the proportions is different from .10.

17. Because $4.168 < 4.400 < 14.684$
   [i.e., $\chi^2_{9,.90} < 4.400 < \chi^2_{9,.10}$],
it must be $.10 < $ P-value $ < .90$ [and the P-value is very close to .90].

19. NOTE: Both outcomes having the same expected frequency is equivalent to $p_1 = p_2 = .5$.

a. H_o: $p_1 = p_2 = .5$
H_1: at least one of the proportions is different from .5
$\alpha = .05$
C.R. $\chi^2 > \chi^2_{1,.05} = 3.841$
calculations

type	O	E	O-E	$(O-E)^2$	$(O-E)^2/E$
A	f_1	$(f_1+f_2)/2$	$(f_1-f_2)/2$	$(f_1-f_2)^2/4$	$[(f_1-f_2)^2/4]/[(f_1+f_2)/2]$
B	f_2	$(f_1+f_2)/2$	$(f_2-f_1)/2$	*$(f_1-f_2)^2/4$	$[(f_1-f_2)^2/4]/[(f_1+f_2)/2]$
	f_1+f_2	f_1+f_2			$[(f_1-f_2)^2/2]/[(f_1+f_2)/2]$

$\chi^2 = \Sigma[(O-E)^2/E]$ *NOTE: $(f_2-f_1)^2 = (f_1-f_2)^2$
$= [(f_1-f_2)^2/2]/[(f_1+f_2)/2]$
$= (f_1-f_2)^2/(f_1+f_2)$

b. H_o: $p = .5$
H_1: $p \neq .5$
$\alpha = .05$
C.R. $z < -z_{.025} = -1.960$
   $z > z_{.025} = 1.960$
calculations:

$z_p = (\hat{p} - \mu_p)/\sigma_p$
$= [f_1/(f_1+f_2) - .5]/\sqrt{(.5)(.5)/(f_1+f_2)}$
$= [.5(f_1-f_2)/(f_1+f_2)]/\sqrt{(.5)(.5)/(f_1+f_2)}$
$= [(f_1-f_2)/(f_1+f_2)]/\sqrt{1/(f_1+f_2)}$
$= (f_1-f_2)/\sqrt{f_1+f_2}$

Note that $z^2 = \chi^2$ since $[(f_1-f_2)/\sqrt{f_1+f_2}]^2 = (f_1-f_2)^2/(f_1+f_2)$ and $(\pm 1.960)^2 = 3.841$.

21. NOTE: Usually the hypothesized p_i's are given and the formula $E_i = np_i$ is used to find the individual expected values. Here, the individual expected values are given (and they sum to $n = 53$) and the formula must be used "in reverse" to solve for the hypothesized p_i's.

H_o: $p_{1,2} = .17$, $p_3 = .15$, $p_4 = .13$, $p_{5,6} = .19$, $p_{7,8} = .13$, $p_{9,10} = .23$

H_1: at least one of the proportions is different from what is claimed

$\alpha = .05$

C.R. $\chi^2 > \chi^2_{5,.05} = 11.071$

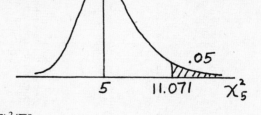

calculations:

slot	O	E	$(O-E)^2/E$
1,2	10	9	.111
3	8	8	.000
4	9	7	.571
5,6	8	10	.400
7,8	3	7	2.286
9,10	15	12	.750
	53	53	4.118

$\chi^2 = \Sigma[(O-E)^2/E]$
$= 4.118$

conclusion:

Do not reject H_o; there is not sufficient evidence to conclude that at least one of the proportions is different from what is claimed.

10-3 Contingency Tables

NOTE: For each row and each column it must be true that $\Sigma O = \Sigma E$. After the marginal row and column totals are calculated, both the row totals and the column totals must sum to produce the same grand total. If either of the preceding is not true, then an error has been made and further calculations have no meaning. In addition, the following are true for all χ^2 contingency table analyses in this manual.

* The E values for each cell are given in parentheses below the O values.
* The addends used to calculate the χ^2 test statistic follow the physical arrangement of the cells in the original contingency table. This practice makes it easier to monitor the large number of inter- mediate steps involved and helps to prevent errors caused by missing or double-counting cells.
* The accompanying chi-square illustration follows the "usual" shape as pictured with Table A-6, even though that shape is not correct for df=1 or df=2.

1. H_o: type of crime and criminal/victim connection are independent

   H_1: type of crime and criminal/victim connection are related

   $\alpha = .05$

   C.R. $\chi^2 > \chi^2_{2,.05} = 5.991$

   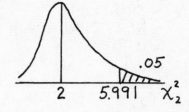

   calculations:

		CRIME			
		H	R	A	
	S	12	379	727	1118
CNCTN		(29.93)	(284.64)	(803.43)	
	A	39	106	642	787
		(21.07)	(200.36)	(565.57)	
		51	485	1369	1905

$\chi^2 = \Sigma[(O-E)^2/E]$
$= 10.7418 + 31.2847 + 7.2715$
$\quad 15.2600 + 44.4425 + 10.3298$
$= 119.30$

conclusion:

Reject H_o; there is sufficient evidence to conclude that the type of crime and the criminal/victim connection are related.

3. H_o: drug treatment and oral reaction are independent
   H_1: drug treatment and oral reaction are related
   $\alpha = .05$
   C.R. $\chi^2 > \chi^2_{1,.05} = 3.841$
   calculations:

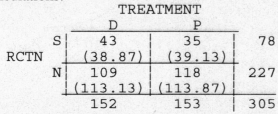

		TREATMENT		
		D	P	
	S	43	35	78
RCTN		(38.87)	(39.13)	
	N	109	118	227
		(113.13)	(113.87)	
		152	153	305

$$\chi^2 = \Sigma[(O-E)^2/E]$$
$$= .4383 + .4355$$
$$\quad .1506 + .1496$$
$$= 1.174$$

conclusion:

Do not reject H_o; there is not sufficient evidence to conclude that the drug treatment and the oral reaction are related.

A person thinking about using Nicorette might still want to be concerned about mouth soreness. While he cannot be 95% sure that there _is_ soreness associated with Nicorette, neither can he be sure that there is _not_ such soreness.

5. H_o: success on the test and group membership are independent
   H_1: success on the test and group membership are related
   $\alpha = .05$
   C.R. $\chi^2 > \chi^2_{1,.05} = 3.841$
   calculations:

		TEST		
		P	F	
	A	10	14	24
GROUP		(17.49)	(6.51)	
	B	417	145	562
		(409.51)	(152.49)	
		427	159	586

$$\chi^2 = \Sigma[(O-E)^2/E]$$
$$= 3.2062 + 8.6105$$
$$\quad .1369 + .3677$$
$$= 12.321$$

conclusion:

Reject H_o; there is sufficient evidence to conclude that success on the test and group membership are related.

NOTE: This does not necessarily mean that anything unfair or illegal is occurring (e.g., that the minority group's tests are graded improperly or that the test is deliberately or unintentionally culturally biased).

7. H_o: wearing a helmet and receiving facial injuries are independent
H_1: wearing a helmet and receiving facial injuries are related
$\alpha = .05$
C.R. $\chi^2 > \chi^2_{1,.05} = 3.841$
calculations:

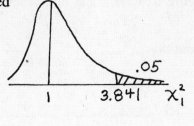

		HELMET		
		Y	N	
INJR	Y	30	182	212
		(45.11)	(166.89)	
	N	83	236	319
		(67.89)	(251.11)	
		113	418	531

$$\chi^2 = \Sigma[(O-E)^2/E]$$
$$= 5.0640 + 1.3690$$
$$\quad 3.3654 + .9098$$
$$= 10.708$$

conclusion:

Reject H_o; there is sufficient evidence to conclude that wearing a helmet and receiving facial injuries are related.

NOTE: There is a statistical relationship, but not necessarily a cause-and-effect relationship. This test does not really address the question "Does a helmet seem to be effective in helping to prevent facial injuries in a crash?" The fact that people who wear helmets receive statistically fewer facial injuries could be due to other factors than the helmet preventing the injury (e.g., people who wear helmets might be safer people who go slower and whose accidents are less serious).

9. H_o: having an accident and using a cellular phone are independent
H_1: having an accident and using a cellular phone are related
$\alpha = .05$
C.R. $\chi^2 > \chi^2_{1,.05} = 3.841$
calculations:

		ACCIDENT		
		Y	N	
PHONE	Y	23	282	305
		(27.76)	(277.24)	
	N	46	407	453
		(41.24)	(411.76)	
		69	689	758

$$\chi^2 = \Sigma[(O-E)^2/E]$$
$$= .8174 + .0819$$
$$\quad .5503 + .0551$$
$$= 1.505$$

conclusion:

Do not reject H_o; there is not sufficient evidence to conclude that having an accident and using a cellular phone are related.

No; based on these results it does not appear that the use of cellular phones affects driving safety.

NOTE: The exercise's statement that the data are from "a study of randomly selected car accidents and drivers who use cellular phones" does not explain how the 407 persons who did neither appear in the survey. Even rejecting the above hypothesis, however, only is evidence that having accidents and having cellular phones are related, not that the phones cause accidents. Any cause-and-effect relationship could even go the other way: drivers tending to have accidents might tend to get cellular phones to be able to call for help.

11. H_o: cause of death and determining agent are independent
H_1: cause of death and determining agent are related
$\alpha = .05$
C.R. $\chi^2 > \chi^2_{3,.05} = 7.815$
calculations:

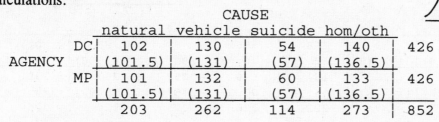

		CAUSE				
		natural	vehicle	suicide	hom/oth	
AGENCY	DC	102 (101.5)	130 (131)	54 (57)	140 (136.5)	426
	MP	101 (101.5)	132 (131)	60 (57)	133 (136.5)	426
		203	262	114	273	852

$$\chi^2 = \Sigma[(O-E)^2/E]$$
$$= .0025 + .0076 + .1579 + .0897$$
$$\quad .0025 + .0076 + .1579 + .0897$$
$$= .515$$

conclusion:
Do not reject H_o; there is not sufficient evidence to conclude that the cause of death and the determining agency are related.

13. H_o: smoking and age are independent
H_1: smoking and age are related
$\alpha = .05$
C.R. $\chi^2 > \chi^2_{3,.05} = 7.815$
calculations:

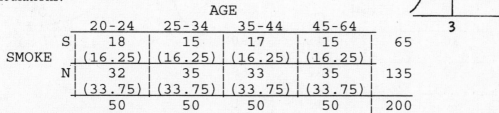

		AGE				
		20-24	25-34	35-44	45-64	
SMOKE	S	18 (16.25)	15 (16.25)	17 (16.25)	15 (16.25)	65
	N	32 (33.75)	35 (33.75)	33 (33.75)	35 (33.75)	135
		50	50	50	50	200

$$\chi^2 = \Sigma[(O-E)^2/E]$$
$$= .1885 + .0963 + .0346 + .0962$$
$$\quad .0907 + .0463 + .0167 + .0463$$
$$= .615$$

conclusion:
Do not reject H_o; there is not sufficient evidence to conclude that smoking and age are related.

At present, there are not a higher percentage of smokers in one age group than in another. Targeting cigarette advertising to younger smokers, however, will reach persons who will be around longer and (if consistently successful) will ultimately raise the percentages in all age groups.

15. H_o: time and location of fatal accidents are independent
H_1: time and location of fatal accidents are related
$\alpha = .10$
C.R. $\chi^2 > \chi^2_{7,.10} = 12.017$
calculations:

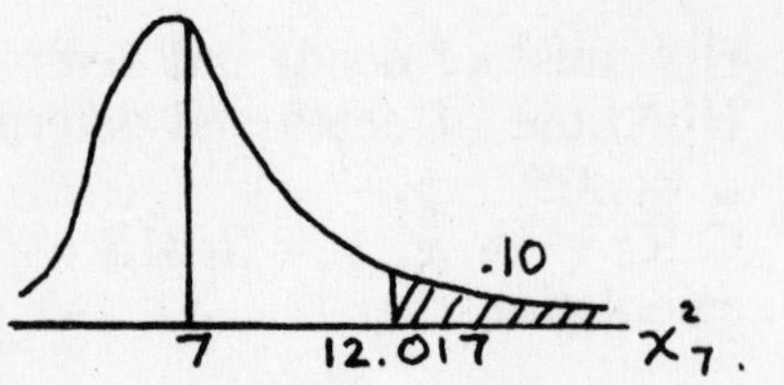

LOC / TIME

	1-4	4-7	7-10	10-1	1-4	4-7	7-10	10-1	
nyc	73 (70.68)	60 (47.58)	53 (51.38)	68 (62.26)	80 (75.03)	67 (88.08)	87 (88.08)	81 (85.91)	569
oth	187 (189.32)	115 (127.42)	136 (137.62)	161 (166.74)	196 (200.97)	257 (235.92)	237 (235.92)	235 (230.09)	1524
	260	175	189	229	276	324	324	326	2093

$\chi^2 = \Sigma[(O-E)^2/E]$
$= .0759 + 3.2448 + .0510 + .5300 + .3288 + 5.0459 + .0133 + .2803$
$.0284 + 1.2115 + .0190 + .1979 + .1228 + 1.8840 + .0050 + .1047$
$= 13.143$

conclusion:
Reject H_o; there is sufficient evidence to conclude that the time and the location of fatal accidents are related.

16. H_o: product opinion and geographic region are independent
H_1: product opinion and geographic region are related
$\alpha = .01$
C.R. $\chi^2 > \chi^2_{4,.01} = 13.277$
calculations:

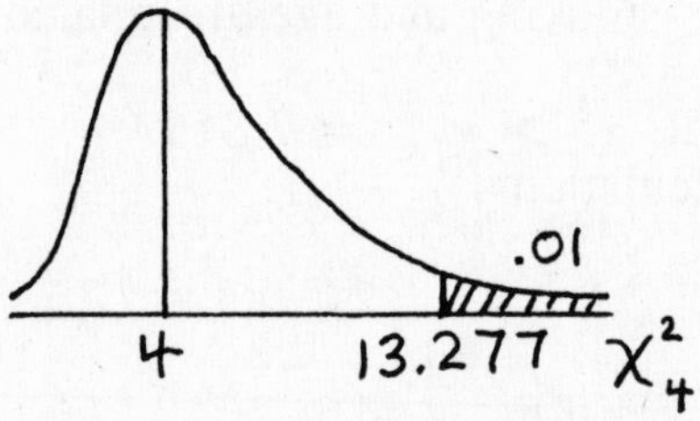

OPINION

REGION	L	D	U	
N	30 (20.43)	15 (26.81)	15 (12.77)	60
S	10 (20.43)	30 (26.81)	20 (12.77)	60
W	40 (39.15)	60 (51.38)	15 (24.47)	115
	80	100	50	235

$\chi^2 = \Sigma[(O-E)^2/E]$
$= 4.4889 + 5.2014 + .3910$
$5.3214 + .3799 + 4.0933$
$.0185 + 1.4461 + 3.6637$
$= 25.009$

conclusion:
Reject H_o; there is sufficient evidence to conclude that product opinion and geographic region are related.
No; based on the above results it would probably not be wise to use the same marketing strategy in each region.

17. Since $25.009 > 14.869$
[i.e., $25.009 > \chi^2_{4,.005}$ -- the largest χ^2 in the row]
then P-value $< .005$.

19. Multiplying each O value by positive constant k multiplies all the totals (and hence each E value) by that same positive constant k. The calculated statistic becomes

$$(\chi^2 \text{ new}) = \Sigma[(kO-kE)^2/kE]$$
$$= \Sigma[k^2(O-E)^2/kE]$$
$$= \Sigma[k(O-E)^2/E]$$
$$= k \cdot \Sigma[(O-E)^2/E]$$
$$= k \cdot (\chi^2 \text{ old})$$

Review Exercises

1. H_o: $p_{Mon} = p_{Tue} = p_{Wed} = \ldots = p_{Sun} = 1/7$
   H_1: at least one of the proportions is different from 1/7
   $\alpha = .05$
   C.R. $\chi^2 > \chi^2_{6,.05} = 12.592$
   calculations:

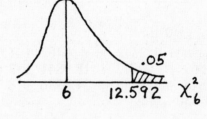

day	O	E	$(O-E)^2/E$
Mon	74	66.286	.8978
Tue	60	66.286	.5961
Wed	66	66.286	.0012
Thu	71	66.286	.3353
Fri	51	66.286	3.5249
Sat	66	66.286	.0012
Sun	76	66.286	1.4236
	464	464.000	6.7801

$\chi^2 = \Sigma[(O-E)^2/E]$
$= 6.780$

conclusion:
   Do not reject H_o; there is not sufficient evidence to conclude that at least one of the proportions is different from 1/7.

3. H_o: $p_{Jan} = p_{Feb} = p_{Mar} = \ldots = p_{Dec} = 1/12$
   H_1: at least one of the proportions is different from 1/12
   $\alpha = .05$
   C.R. $\chi^2 > \chi^2_{11,.05} = 19.675$
   calculations:

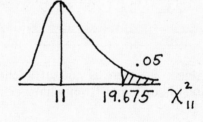

month	O	E	$(O-E)^2/E$
Jan	8	10	.400
Feb	12	10	.400
Mar	9	10	.100
Apr	15	10	2.500
May	6	10	1.600
Jun	12	10	.400
Jul	4	10	3.600
Aug	7	10	.900
Sep	11	10	.100
Oct	11	10	.100
Nov	5	10	2.500
Dec	20	10	10.000
	120	120	22.600

$\chi^2 = \Sigma[(O-E)^2/E]$
$= 22.600$

conclusion:
   Reject H_o; there is sufficient evidence to conclude that at least one of the proportions is different from 1/12.
   No; the fact that some months were offered as answers more than other months says nothing about how often the offered answers were correct and whether or not the subjects have ESP.

5. H_o: county of violation and type of violation are independent
   H_1: county of violation and type of violation are related
   $\alpha = .05$
   C.R. $\chi^2 > \chi^2_{6,.05} = 12.592$
   calculations:

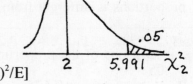

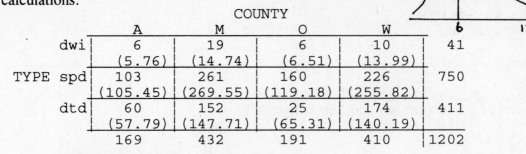

		A	M	O	W	
	dwi	6	19	6	10	41
		(5.76)	(14.74)	(6.51)	(13.99)	
TYPE	spd	103	261	160	226	750
		(105.45)	(269.55)	(119.18)	(255.82)	
	dtd	60	152	25	174	411
		(57.79)	(147.71)	(65.31)	(140.19)	
		169	432	191	410	1202

COUNTY

$$\chi^2 = \Sigma[(O-E)^2/E]$$
$$= .0096 + 1.2342 + .0407 + 1.1355$$
$$.0569 + .2712 + 13.9841 + 3.4768$$
$$.0848 + .1244 + 24.8786 + 8.1533$$
$$= 53.450$$

conclusion:
   Reject H_o; there is sufficient evidence to conclude that the county of the violation and the type of the violation are related.

7. H_o: $p_0 = .80$, $p_1 = .16$, $p_{more} = .04$
   H_1: at least one of the proportions is different from what is claimed
   $\alpha = .05$
   C.R. $\chi^2 > \chi^2_{2,.05} = 5.991$
   calculations:

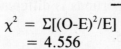

count	O	E	$(O-E)^2/E$
0	172	160	.9000
1	23	32	2.5312
more	5	8	1.1250
	200	200	4.5562

$$\chi^2 = \Sigma[(O-E)^2/E]$$
$$= 4.556$$

conclusion:
   Do not reject H_o; there is not sufficient evidence to conclude that at least one of the proportions is different from what is claimed.

Chapter 11

Analysis of Variance

11-2 One-Way ANOVA with Equal Sample Sizes

NOTE: This section is calculation-oriented. Do not get so involved with the formulas that you miss concepts. This manual arranges the calculations to promote both computational efficiency and understanding of the underlying principles. The following notation is used in this section.

k = the number of groups
n = the number of scores in each group
$\bar{x}_i$ = the mean of group i (where i = 1,2,...,k)
s_i^2 = the variance of group i (where i = 1,2,...,k)
$\bar{\bar{x}}$ = $\Sigma\bar{x}_i/k$ = the mean of the group means
     = the overall mean of all the scores in all the groups
$s_{\bar{x}}^2$ = $\Sigma(\bar{x}_i-\bar{\bar{x}})^2/(k-1)$ = the variance of the group means
s_p^2 = $\Sigma s_i^2/k$ = the mean of the group variances

* Exercise 1 is worked in complete detail, showing even work done on the calculator without having to be written down. Subsequent exercises are worked showing all intermediate steps, but without writing down detail for routine work done on the calculator.

* While the manual typically shows only three decimal places for the intermediate steps, all decimal places were carried in the calculator. DO NOT ROUND OFF INTERMEDIATE ANSWERS. SAVE calculated values that will be used again. See your instructor or class assistant if you need help using your calculator accurately and efficiently.

1. The following preliminary values are identified and/or calculated.
   $k = 3$ numerator df = k-1 = 2
   $n = 5$ denominator df = k(n-1) = 3(4) = 12
   $\bar{\bar{x}}$ = $\Sigma\bar{x}_i/k$ = (98.940 + 98.580 + 97.800)/3
           = 98.440
   $s_{\bar{x}}^2$ = $\Sigma(\bar{x}_i-\bar{\bar{x}})^2/(k-1)$ = [(98.940-98.440)^2 + (98.580-98.440)^2 + (97.800-98.400)^2]/2
              = [(.500)^2 + (.140)^2 + (-.640)^2]/2
              = .6792/2
              = .3396
   s_p^2 = $\Sigma s_i^2/k$ = [(.568)^2 + (.701)^2 + (.752)^2]/3
           = 1.379529/3
           = .459843

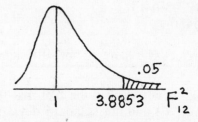

H_o: $\mu_1 = \mu_2 = \mu_3$
H_1: at least one mean is different
$\alpha = .05$
C.R. F > $F_{12,.05}^2$ = 3.8853
calculations:
   F = $ns_{\bar{x}}^2/s_p^2$
     = 5(.3396)/.459843
     = 3.6926
conclusion:
   Do not reject H_o; there is not sufficient evidence to conclude that at least one mean is different.

NOTE: As previously indicated, this manual shows only the intermediate results and not all the routine mathematical calculations for the remaining exercises in this section. Refer to the detail given in exercise 1 for further insights. In addition, each of the remaining exercises in this section introduces a "new twist" to the basic problem (i.e., as presented in exercise 1). These are identified and discussed in a NOTE at the end of each of the exercises.

3. The following preliminary values are identified and/or calculated.

$k = 4$

$n = 62$

$s_{\bar{x}}^2 = \Sigma(\bar{x}_i - \bar{\bar{x}})^2/(k-1)$
$= 36.636/3$
$= 12.212$

$\bar{\bar{x}} = \Sigma\bar{x}_i/k$
$= 4.327$

$s_p^2 = \Sigma s_i^2/k$
$= 29.356/4$
$= 7.339$

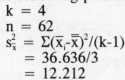

$H_o: \mu_1 = \mu_2 = \mu_3 = \mu_4$
$H_1:$ at least one mean is different
$\alpha = .05$
C.R. $F > F_{244,.05}^3 = 2.6802$
calculations:
$F = ns_{\bar{x}}^2/s_p^2$
$= 62(12.212)/7.339$
$= 103.1651$

conclusion:
  Reject H_o; there is sufficient evidence to conclude that at least one mean is different.

NOTE: The "new twist" is the fact that the accumulation of relatively large sample sizes for several groups pushes the denominator degrees of freedom [df = k(n-1)] past 120, the highest finite entry in the F table. When this occurs, we follow the pattern of using the closest entry and [since 120 is closer to any finite number than ∞ (i.e., infinity)] choose the entry for df = 120.

5. The following preliminary values are identified and/or calculated.

$k = 5$

$n = 10$

$s_{\bar{x}}^2 = \Sigma(\bar{x}_i - \bar{\bar{x}})^2/(k-1)$
$= 93.20/4$
$= 23.30$

$\bar{\bar{x}} = \Sigma\bar{x}_i/k$
$= 102.4$

$s_p^2 = \Sigma s_i^2/k$
$= 755/5$
$= 151$

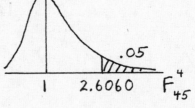

$H_o: \mu_1 = \mu_2 = \mu_3 = \mu_4 = \mu_5$
$H_1:$ at least one mean is different
$\alpha = .05$
C.R. $F > F_{45,.05}^4 = 2.6060$
calculations:
$F = ns_{\bar{x}}^2/s_p^2$
$= 10(23.30)/151$
$= 1.5430$

conclusion:
  Do not reject H_o; there is not sufficient evidence to conclude that at least one mean is different.

NOTE: The "new twist" is the fact that the variability for each group was given in terms of variance (i.e., s^2) and not standard deviation (i.e., s). This means that the given values do not have to be squared again when finding $s_p^2 = \Sigma s_i^2/k$. In problems for which the summary statistics are not given (and for which the reader must make all calculations from the raw data), calculating s^2 instead of s (since it is s^2 that is utilized in subsequent calculations) avoids extra work and the possible introduction of round-off errors due to taking the square root.

7. The following preliminary values are identified and/or calculated.

	A	B	C	D	E
n	4	4	4	4	4
Σx	72	76	76	84	68
Σx^2	1318	1478	1448	1770	1170
$\bar{x}$	18	19	19	21	17
s^2	7.333	11.333	1.333	2.000	4.667

$k = 5$

$n = 4$

$s_{\bar{x}}^2 = \Sigma(\bar{x}_i - \bar{\bar{x}})^2/(k-1)$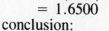

$\quad = 8.80/4$

$\quad = 2.20$

$\bar{\bar{x}} = \Sigma\bar{x}_i/k$

$\quad = 18.8$

$s_p^2 = \Sigma s_i^2/k$

$\quad = 26.667/5$

$\quad = 5.333$

$H_o: \mu_A = \mu_B = \mu_C = \mu_D = \mu_E$

H_1: at least one mean is different

$\alpha = .05$

C.R. $F > F_{15,.05}^4 = 3.0556$

calculations:

$F = ns_{\bar{x}}^2/s_p^2$

$\quad = 4(2.20)/5.333$

$\quad = 1.6500$

conclusion:

Do not reject H_o; there is not sufficient evidence to conclude that at least one mean is different.

NOTE: The "new twist" is the fact that the group means and variances are not given and need to be calculated from the raw data.

9. The following preliminary values are identified and/or calculated.

	1	4	7
n	10	10	10
Σx	1493	1445	1387
Σx^2	225,541	236,287	210,189
$\bar{x}$	149.3	144.5	138.7
s^2	292.900	3053.833	1979.122

$k = 3$

$n = 10$

$s_{\bar{x}}^2 = \Sigma(\bar{x}_i - \bar{\bar{x}})^2/(k-1)$

$\quad = 56.347/2$

$\quad = 28.173$

$\bar{\bar{x}} = \Sigma\bar{x}_i/k$

$\quad = 144.167$

$s_p^2 = \Sigma s_i^2/k$

$\quad = 5325.856/3$

$\quad = 1775.285$

$H_o: \mu_1 = \mu_4 = \mu_7$

H_1: at least one mean is different

$\alpha = .05$

C.R. $F > F_{27,.05}^2 = 3.3541$

calculations:

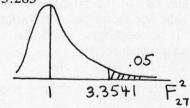

$F = ns_{\bar{x}}^2/s_p^2$

$\quad = 10(28.173)/1775.185$

$\quad = .1587$

conclusion:

Do not reject H_o; there is not sufficient evidence to conclude that at least one mean is different.

No, no zone seems to have homes with higher (or lower) selling prices.

NOTE: The "new twist" is the fact that the selling prices (whose overall average is $144,167) can be analyzed in terms of 1000's of dollars instead of dollars. This is so because dividing all the selling prices by 1000 also divides all the group means by 1000. The variances (of the sample means and the pooled within group estimate) in the numerator and the denominator of the F statistic are each divided by 1000^2, and the F statistic is precisely what it would have been had we worked with dollars instead of $1000's.

11. The following preliminary values are identified and/or calculated.

	1	4	7
n	10	10	10
Σx	8.65	5.71	21.39
Σx^2	9.4507	5.0457	78.9271
$\bar{x}$	.865	.571	2.139
s^2	.219	.198	3.686

$k = 3$

$n = 10$

$s_{\bar{x}}^2 = \Sigma(\bar{x}_i - \bar{\bar{x}})^2/(k-1)$

$\qquad = 1.389/2$

$\qquad = .695$

$\bar{\bar{x}} = \Sigma\bar{x}_i/k$

$\qquad = 1.192$

$s_p^2 = \Sigma s_i^2/k$

$\qquad = 4.103/3$

$\qquad = 1.368$

$H_o: \mu_1 = \mu_4 = \mu_7$

$H_1:$ at least one mean is different

$\alpha = .05$

C.R. $F > F_{27,.05}^2 = 3.3541$

calculations:

$\quad F = ns_{\bar{x}}^2/s_p^2$

$\qquad = 10(.695)/1.368$

$\qquad = 5.0793$

conclusion:

Reject H_o; there is sufficient evidence to conclude that at least one mean is different.
Yes, one zone (viz., zone 7) does appear to have larger lots.

NOTE: The "new twist" is the fact that the author expects a specific deduction on the basis of rejecting the null hypothesis. In truth, the only conclusion possible from the ANOVA test is the one stated, that "at least one mean is different." Considering the ordering of the sample means, there are three distinct possibilities: $\mu_7 > \mu_1 > \mu_4$, $\mu_7 > \mu_1 = \mu_4$, $\mu_7 = \mu_1 > \mu_4$. There are further statistical procedures that can be employed to determine which of the preceding cases should be concluded, but they are beyond the scope of the text.

NOTE ALSO: One of the assumptions of the ANOVA test is that all the groups have the same variance (and that common variance is estimated by s_p^2 -- the average of the group variances). In this exercise, one of the sample variances is almost 20 times larger than the others, and it is questionable whether the population variances are truly equal. This means that perhaps the ANOVA test is not appropriate and that a professional statistician should be consulted about a more advanced procedure.

13. a. $n_1 > 30$ and $n_2 > 30$, use z (with s's for σ's)]

$\bar{x}_1 - \bar{x}_2 = 197 - 202 = -5$

H_o: $\mu_1 - \mu_2 = 0$

H_1: $\mu_1 - \mu_2 \neq 0$

$\alpha = .05$

C.R. $z < -z_{.025} = -1.960$
$\quad z > z_{.025} = 1.960$

calculations:

$z_{\bar{x}_1 - \bar{x}_2} = (\bar{x}_1 - \bar{x}_2 - \mu_{\bar{x}_1 - \bar{x}_2})/\sigma_{\bar{x}_1 - \bar{x}_2}$
$\quad = (-5 - 0)/\sqrt{(18)^2/50 + (20)^2/50}$
$\quad = -5/3.805$
$\quad = -1.314$

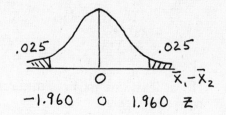

conclusion:

Do not reject H_o; there is not sufficient evidence to conclude that $\mu_1 - \mu_2 \neq 0$.

b. $n_2 > 30$ and $n_3 > 30$, use z (with s's for σ's)]

$\bar{x}_2 - \bar{x}_3 = 202 - 208 = -6$

H_o: $\mu_2 - \mu_3 = 0$

H_1: $\mu_2 - \mu_3 \neq 0$

$\alpha = .05$

C.R. $z < -z_{.025} = -1.960$
$\quad z > z_{.025} = 1.960$

calculations:

$z_{\bar{x}_2 - \bar{x}_3} = (\bar{x}_2 - \bar{x}_3 - \mu_{\bar{x}_2 - \bar{x}_3})/\sigma_{\bar{x}_2 - \bar{x}_3}$
$\quad = (-6 - 0)/\sqrt{(20)^2/50 + (23)^2/50}$
$\quad = -6/4.310$
$\quad = -1.392$

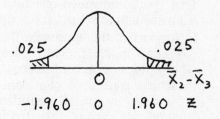

conclusion:

Do not reject H_o; there is not sufficient evidence to conclude that $\mu_1 - \mu_2 \neq 0$.

c. $n_1 > 30$ and $n_3 > 30$, use z (with s's for σ's)]

$\bar{x}_1 - \bar{x}_3 = 197 - 208 = -11$

H_o: $\mu_1 - \mu_3 = 0$

H_1: $\mu_1 - \mu_3 \neq 0$

$\alpha = .05$

C.R. $z < -z_{.025} = -1.960$
$\quad z > z_{.025} = 1.960$

calculations:

$z_{\bar{x}_1 - \bar{x}_3} = (\bar{x}_1 - \bar{x}_3 - \mu_{\bar{x}_1 - \bar{x}_3})/\sigma_{\bar{x}_1 - \bar{x}_3}$
$\quad = (-11 - 0)/\sqrt{(18)^2/50 + (23)^2/50}$
$\quad = -11/4.130$
$\quad = -2.663$

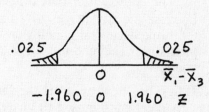

conclusion:

Reject H_o; there is sufficient evidence to conclude that $\mu_1 - \mu_2 \neq 0$ (in fact, $\mu_1 - \mu_3 < 0$).

d. The following preliminary values are identified and/or calculated.

$k = 3$

$n = 50$

$s_{\bar{x}}^2 = \Sigma(\bar{x}_i - \bar{\bar{x}})^2/(k-1)$
$= 60.667/2$
$= 30.333$

$\bar{\bar{x}} = \Sigma\bar{x}_i/k$
$= 202.333$

$s_p^2 = \Sigma s_i^2/k$
$= 1253/3$
$= 417.667$

H_o: $\mu_1 = \mu_2 = \mu_3$
H_1: at least one mean is different
$\alpha = .05$
C.R. $F > F_{147, .05}^2 = 3.0718$
calculations:

$F = ns_{\bar{x}}^2/s_p^2$
$= 50(30.333)/417.667$
$= 3.6313$

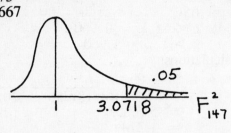

conclusion:

Reject H_o; there is sufficient evidence to conclude that at least one mean is different.

e. This is a complicated question whose complete answer requires more than can be ascertained from this exercise and the scope of this text. Each approach has its advantages.

In support of the approach of part (d), it may be said that this approach
* seems more efficient, since it requires only a single test.
* provides 95% confidence. The approach of parts (a),(b),(c) involves three tests, each of which gives 95% confidence separately -- but 95% confidence in three different tests leads to an overall confidence less than 95%. [For independent tests (which these are not), for example, the overall confidence rate would be $(.95)^3 = .857 = 85.7\%$.

In support of the approach of parts (a),(b),(c), it may be said that this approach
* can use either one-tailed or two-tailed tests.
* can identify precisely which means are significantly different and which are not, while the approach of part (d) can only say that "at least one mean is different."
* seems more certain. The -2.663 of part (c) was farther into the critical region determined by -1.960 than the 3.6313 of part (d) was into the critical region determined by 3.0718. For part (c), P-value = 2(.0039) = .0078; for part (d), .01 < P-value < .025. The approach of part (c) seems to give stronger evidence of a difference.

Because the advantages given above for the approach of part (d) are significant, and because the limited general conclusion of the approach of part (d) can be improved by additional procedures (beyond the scope of the text), that approach is generally considered the better one.

NOTE: The "new twist" is the fact that two different methods (those of chapter 8 and those of this chapter) can be used to analyze the same problem and that these different methods each have their own advantages and disadvantages.

15. a. Adding a constant to each of the sample means (because, it is assumed, that constant was added to each of the individual scores) does not change the calculated $F = ns_{\bar{x}}^2/s_p^2$. Since all the means increase by the same amount, the variance among them (i.e., $s_{\bar{x}}^2$) is not affected; since all the individual scores increase by the same amount, the group variances and hence the average group variance (i.e., s_p^2) is not affected.

   b. This question has two possible interpretations:

   * Multiplying each sample mean by a constant (because, it is assumed, each individual scores was multiplied by that constant) does not change the calculated $F = ns_{\bar{x}}^2/s_p^2$. Since all the means are multiplied by a constant, the variance among them (i.e., $s_{\bar{x}}^2$) is multiplied by the square of that constant; since all the individual scores are multiplied by a constant, the group variances and hence the average group variance (i.e., s_p^2) is multiplied by the square of that constant. Since both the numerator and the denominator are multiplied by the same amount, the calculated F is not affected.

   * Multiplying each sample mean by a constant (because, it is assumed, we were drawing from populations whose means were a constant multiple of the original means but whose variances were the same as the original variances) multiples the calculated $F = ns_{\bar{x}}^2/s_p^2$ by the square of that constant. Since all the means are multiplied by a constant, the variance among them (i.e., $s_{\bar{x}}^2$) is multiplied by the square of that constant; since variance of the individual scores is not changed, neither is s_p^2. Since the numerator is multiplied by the square of that constant and the denominator is not affected, the calculated F is multiplied by the square of that constant.

   c. If the 5 samples have the same mean, then there is no variability among the means and $s_{\bar{x}}^2 = 0$. Assuming there was variability among the individual scores, $s_p^2 \neq 0$. Since the numerator is 0 and the denominator is not, the calculated F statistic equals 0.

NOTE: The "new twist" is the fact that questions are posed without specific numerical data. This tests both the reader's ability to reason abstractly and his understanding of the underlying statistical principles.

11-3 One-Way ANOVA with Unequal Sample Sizes

NOTE: This section is calculation-oriented. Do not get so involved with the formulas that you miss concepts. This manual arranges the calculations to promote both computational efficiency and understanding of the underlying principles. The following notation is used in this section.

k = the number of groups

n_i = the number of scores in group i (where i = 1,2,...k)

$\bar{x}_i$ = the mean of group i (where i = 1,2,...,k)

s_i^2 = the variance of group i (where i = 1,2,...,k)

The following items are calculated as in the previous section except that the unequal sample sizes require a weighting using either n_i or $(n_i-1) = df_i$.

$\bar{\bar{x}} = \Sigma\bar{x}_i/k$ becomes $\Sigma n_i\bar{x}_i/\Sigma n_i$

        = the overall weighted mean of all the scores in all the groups

$ns_{\bar{x}}^2 = n\Sigma(\bar{x}_i-\bar{\bar{x}})^2/(k-1)$ becomes $\Sigma n_i(\bar{x}_i-\bar{\bar{x}})^2/(k-1)$

        = the weighted variance estimate from the group means

$s_p^2 = \Sigma s_i^2/k$ becomes $\Sigma(n_i-1)s_i^2/\Sigma(n_i-1) = \Sigma df_i s_i^2/\Sigma df_i$

        = the weighted mean of the group variances

* All exercises are worked in complete detail, with the work arranged in an ANOVA table. In addition, NOTES are given commenting on the order in which the values in the table were filled in and on how new values were obtained from previous ones.

* While the manual typically shows only a certain number of decimal places (depending on the exercise) for the intermediate steps, all decimal places were carried in the calculator. DO NOT ROUND OFF INTERMEDIATE ANSWERS. SAVE calculated values that will be used again. See your instructor or class assistant if you need help using your calculator accurately and efficiently.

* Use of the ANOVA table to organize the calculations is strongly recommended. The calculated values appear in the ANOVA table as follows.

source	SS	df	MS	F
treatment	$\Sigma n_i(\bar{x}_i-\bar{\bar{x}})^2$	$k-1$	SS_{Trt}/df_{Trt}	MS_{Trt}/MS_E
error	$\Sigma df_i s_i^2$	$\Sigma df_i = \Sigma n_i - k$	SS_E/df_E	
total	$\Sigma(x-\bar{\bar{x}})^2$	$\Sigma n_i - 1$		

The SS and df columns are additive as indicated. If $MS_{Tot} = \Sigma(x-\bar{\bar{x}})^2/(\Sigma n_i-1)$ is calculated, it gives s^2 (i.e., the variance for all the scores considered as one big group) and not the sum of the MS column.

1. The following preliminary values are identified and/or calculated.

	G	F	P	total
n	5	7	6	18
Σx	437	484	357	1278
Σx^2	38289	34168	21469	93926
$\bar{x}$	87.400	69.143	59.500	71.000
s^2	23.800	117.143	45.500	

$\bar{\bar{x}} = \Sigma n_i \bar{x}_i / \Sigma n_i$

$\quad = [5(87.400) + 7(69.143) + 6(59.500)]/(5 + 7 + 6)$

$\quad = 1278/18$

$\quad = 71.000$ [NOTE: This must always agree with the $\bar{x}$ in the "total" column.]

$\Sigma n_i(\bar{x}_i - \bar{\bar{x}})^2 = 5(87.400 - 71.000)^2 + 7(69.143 - 71.000)^2 + 6(59.500 - 71.000)^2$

$\quad\quad\quad = 2162.443$

$\Sigma df_i s_i^2 = 4(23.800) + 6(117.142) + 5(45.500)$

$\quad\quad\quad = 1025.558$

$H_o: \mu_G = \mu_F = \mu_P$

$H_1:$ at least one mean is different

$\alpha = .05$

C.R. $F > F^2_{15,.05} = 3.6823$

calculations:

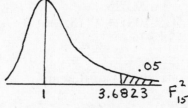

$F = MS_{Trt}/MS_E$

$\quad = 1081.221/68.371$

$\quad = 15.8141$

source	SS	df	MS	F
Trt	2162.443	2	1081.221	15.8141
Error	1025.558	15	68.371	
Total	3188.000	17		

conclusion:

Reject H_o; there is sufficient evidence to conclude that at least one mean is different.

NOTE: Complete the ANOVA table as follows:

(1) Enter $SS_{Trt} = \Sigma n_i(\bar{x}_i - \bar{\bar{x}})^2$ and $SS_E = \Sigma df_i s_i^2$ values from the preliminary calculations.

(2) Enter $df_{Trt} = k-1$ and $df_E = \Sigma df_i = \Sigma(n_i - 1) = \Sigma n_i - k$.

(3) Add the SS and df columns to find SS_{Tot} and df_{Tot}. [The df_{Tot} must equal $\Sigma n_i - 1$.]

(4) Calculate $MS_{Trt} = SS_{Trt}/df_{Trt}$ and $MS_E = SS_E/df_E$.

(5) Calculate $F = MS_{Trt}/MS_E$.

As a final check, calculate s^2 (i.e., the variance of all the scores in one large group) two different ways as indicated below. If these answers agree, the problem is probably correct.

* from the "total" column in the table for the preliminary calculations:

$\quad s^2 = [n\Sigma x^2 - (\Sigma x)^2]/[n(n-1)]$

$\quad\quad = [18(93926) - (1278)^2]/[18(17)]$

$\quad\quad = 57384/306$

$\quad\quad = 187.53$

* from the "total" row of the ANOVA table

$\quad s^2 = SS_{Tot}/df_{Tot}$

$\quad\quad = 3188.000/17$

$\quad\quad = 187.53$

3. The following preliminary values are identified and/or calculated.

	A	B	C	D	total
n	6	10	8	5	29
Σx	93	74	147	92	406
Σx^2	1537	684	2951	1710	6882
$\bar{x}$	15.500	7.400	18.375	18.400	14.000
s^2	19.100	15.156	35.696	4.300	

$\bar{\bar{x}} = \Sigma n_i \bar{x}_i / \Sigma n_i$

$\quad = [6(15.500) + 10(7.400) + 8(18.375) + 5(18.400)]/(6 + 10 + 8 + 5)$

$\quad = 406/29$

$\quad = 14.000$ [NOTE: This must always agree with the $\bar{x}$ in the "total" column.]

$\Sigma n_i(\bar{x}_i - \bar{\bar{x}})^2 = 6(15.500 - 14.000)^2 + 10(7.400 - 14.000)^2 + 8(18.375 - 14.000)^2$
$\qquad\qquad\qquad\qquad + 5(18.400 - 14.000)^2$

$\qquad\qquad = 699.025$

$\Sigma df_i s_i^2 = 5(19.100) + 9(15.156) + 7(35.696) + 4(4.300)$

$\qquad = 498.975$

$H_o: \mu_A = \mu_B = \mu_C = \mu_D$
$H_1:$ at least one mean is different
$\alpha = .01$
C.R. $F > F_{25,.01}^3 = 4.6755$
calculations:

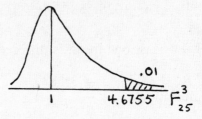

source	SS	df	MS	F
Trt	699.025	3	233.008	11.6743
Error	498.975	25	19.959	
Total	1198.000	28		

$F = MS_{Trt}/MS_E$

$\quad = 233.008/19.959$

$\quad = 11.6743$

conclusion:

Reject H_o; there is sufficient evidence to conclude that at least one mean is different.

NOTE: Complete the ANOVA table as follows:

(1) Enter $SS_{Trt} = \Sigma n_i(\bar{x}_i - \bar{\bar{x}})^2$ and $SS_E = \Sigma df_i s_i^2$ values from the preliminary calculations.

(2) Enter $df_{Trt} = k-1$ and $df_E = \Sigma df_i = \Sigma(n_i-1) = \Sigma n_i - k$.

(3) Add the SS and df columns to find SS_{Tot} and df_{Tot}. [The df_{Tot} must equal $\Sigma n_i - 1$.]

(4) Calculate $MS_{Trt} = SS_{Trt}/df_{Trt}$ and $MS_E = SS_E/df_E$.

(5) Calculate $F = MS_{Trt}/MS_E$.

As a final check, calculate s^2 (i.e., the variance of all the scores in one large group) two different ways as indicated below. If these answers agree, the problem is probably correct.

* from the "total" column in the table for the preliminary calculations:

$\quad s^2 = [n\Sigma x^2 - (\Sigma x)^2]/[n(n-1)]$

$\qquad = [29(6882) - (406)^2]/[29(28)]$

$\qquad = 34742/812$

$\qquad = 42.79$

* from the "total" row of the ANOVA table

$\quad s^2 = SS_{Tot}/df_{Tot}$

$\qquad = 1198.000/28$

$\qquad = 42.79$

5. The following preliminary values are identified and/or calculated.

	1	2	3	4	5	total
n	11	11	10	9	7	48
Σx	36.7	39.6	32.5	27.6	25.5	161.9
Σx^2	124.49	144.68	106.73	85.22	94.65	555.77
$\bar{x}$	3.3364	3.6000	3.2500	3.0667	3.6429	3.3729
s^2	.2045	.2120	.1228	.0725	.2929	

$\bar{\bar{x}} = \Sigma n_i \bar{x}_i / \Sigma n_i$

$= [11(3.3364) + 11(3.6000) + 10(3.2500) + 9(3.0667) + 7(3.6429)]/48$

$= 161.9/48$

$= 3.3729$ [NOTE: This must always agree with the $\bar{x}$ in the "total" column.]

$\Sigma n_i(\bar{x}_i - \bar{\bar{x}})^2 = 11(3.3364 - 3.3729)^2 + 11(3.6000 - 3.3729)^2 + 10(3.2500 - 3.3729)^2$
$\qquad\qquad\qquad + 9(3.0667 - 3.3729)^2 + 7(3.6429 - 3.3729)^2$

$\qquad\qquad = 2.087$

$\Sigma df_i s_i^2 = 10(.2045) + 10(.2120) + 9(.1228) + 8(.0725) + 6(.2929)$

$\qquad = 7.608$

H_o: $\mu_1 = \mu_2 = \mu_3 = \mu_4 = \mu_5$

H_1: at least one mean is different

$\alpha = .05$

C.R. $F > F_{43,.05}^4 = 2.6060$

calculations:

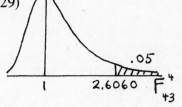

$F = MS_{Trt}/MS_E$

$= .522/.177$

$= 2.9491$

source	SS	df	MS	F
Trt	2.087	4	.522	2.9491
Error	7.608	43	.177	
Total	9.694	47		

conclusion:

Reject H_o; there is sufficient evidence to conclude that at least one mean is different.

NOTE: Complete the ANOVA table as follows:

(1) Enter $SS_{Trt} = \Sigma n_i(\bar{x}_i - \bar{\bar{x}})^2$ and $SS_E = \Sigma df_i s_i^2$ values from the preliminary calculations.

(2) Enter $df_{Trt} = k-1$ and $df_E = \Sigma df_i = \Sigma(n_i - 1) = \Sigma n_i - k$.

(3) Add the SS and df columns to find SS_{Tot} and df_{Tot}. [The df_{Tot} must equal $\Sigma n_i - 1$.]

(4) Calculate $MS_{Trt} = SS_{Trt}/df_{Trt}$ and $MS_E = SS_E/df_E$.

(5) Calculate $F = MS_{Trt}/MS_E$.

As a final check, calculate s^2 (i.e., the variance of all the scores in one large group) two different ways as indicated below. If these answers agree, the problem is probably correct.

* from the "total" column in the table for the preliminary calculations:

$s^2 = [n\Sigma x^2 - (\Sigma x)^2]/[n(n-1)]$

$= [48(555.77) - (161.9)^2]/[48(47)]$

$= 465.35/2256$

$= .206$

* from the "total" row of the ANOVA table

$s^2 = SS_{Tot}/df_{Tot}$

$= 9.694/47$

$= .206$

7. The following preliminary values are identified and/or calculated.

	R	O	Y	B	T	G	total
n	17	9	24	30	11	9	100
Σx	15.373	8.267	21.939	27.769	10.242	8.011	91.601
Σx^2	13.929193	7.596853	20.085206	25.782618	9.551278	7.147987	84.093135
$\bar{x}$	.90429	.91856	.91413	.92563	.93109	.89011	.91601
s^2	.001717	.000394	.001314	.002714	.001504	.002163	

$$\bar{\bar{x}} = \Sigma n_i \bar{x}_i / \Sigma n_i$$
$$= [17(.90429) + 9(.91856) + 24(.91413) + 30(.92562) + 11(.93109) + 9(.89011)]/100$$
$$= 91.601/100$$
$$= .91601 \quad \text{[NOTE: This must always agree with the } \bar{x} \text{ in the "total" column.]}$$
$$\Sigma n_i(\bar{x}_i - \bar{\bar{x}})^2 = 17(.90429-.91601)^2 + 9(.91856-.91601)^2 + 24(.91413-.91601)^2$$
$$+ 30(.92563-.91601)^2 + 11(.93109-.91601)^2 + 9(.89011-.91601)^2$$
$$= .01379$$
$$\Sigma df_i s_i^2 = 16(.001717) + 8(.000394) + 23(.001314)$$
$$+ 29(.002714) + 10(.001504) + 8(.002163)$$
$$= .17190$$

H_o: $\mu_R = \mu_O = \mu_Y = \mu_B = \mu_T = \mu_G$
H_1: at least one mean is different
$\alpha = .05$
C.R. $F > F^5_{94,.05} = 2.2899$
calculations:

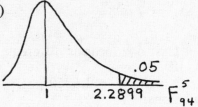

$F = MS_{Trt}/MS_E$
$\quad = .00276/.00183$
$\quad = 1.5082$

source	SS	df	MS	F
Trt	.01379	5	.00276	1.5082
Error	.17190	94	.00183	
Total	.18569	99		

conclusion:

Do not reject H_o; there is not sufficient evidence to conclude that at least one mean is different.

No; we cannot be 95% certain that the company has a problem that requires corrective action.

NOTE: Complete the ANOVA table as follows:
(1) Enter $SS_{Trt} = \Sigma n_i(\bar{x}_i - \bar{\bar{x}})^2$ and $SS_E = \Sigma df_i s_i^2$ values from the preliminary calculations.
(2) Enter $df_{Trt} = k-1$ and $df_E = \Sigma df_i = \Sigma(n_i-1) = \Sigma n_i - k$.
(3) Add the SS and df columns to find SS_{Tot} and df_{Tot}. [The df_{Tot} must equal $\Sigma n_i - 1$.]
(4) Calculate $MS_{Trt} = SS_{Trt}/df_{Trt}$ and $MS_E = SS_E/df_E$.
(5) Calculate $F = MS_{Trt}/MS_E$.
As a final check, calculate s^2 (i.e., the variance of all the scores in one large group) two different ways as indicated below. If these answers agree, the problem is probably correct.
* from the "total" column in the table for the preliminary calculations:
$$s^2 = [n\Sigma x^2 - (\Sigma x)^2]/[n(n-1)]$$
$$= [100(84.093135) - (91.601)^2]/[100(99)]$$
$$= 18.571799/9900$$
$$= .001876$$
* from the "total" row of the ANOVA table
$$s^2 = SS_{Tot}/df_{Tot}$$
$$= .18569/99$$
$$= .001876$$

9. The completed ANOVA table appears below.

source	SS	df	MS	F
Trt	2.17	2	1.085	.1542
Error	112.57	16	7.036	
Total	114.74	18		

Complete the ANOVA table as follows.
(1) Since SS_{Trt} + 112.57 = 114.74, it follows that SS_{Trt} = 114.74 - 112.57 = 2.17.
(2) Since there were three groups, k = 3 and df_{Trt} = k - 1 = 2.
(3) Since n_1 = 5 and n_2 = 7 and n_3 = 7, df_E = Σdf_i = 4 + 6 + 6 = 16.
(4) MS_{Trt} = SS_{Trt}/df_{Trt} = 2.17/2 = 1.085.
(5) MS_E = SS_E/df_E = 112.57/16 = 7.036. [STORE this, do not round it off.]
(6) F = MS_{Trt}/MS_E = 1.085/7.036 = .1542

NOTE: The text could also have asked, "What would be the variance of all 19 scores taken as one large group?" The answer is s^2 = SS_{Tot}/df_{Tot} = 114.74/18 = 6.374.

11. The following preliminary values are identified and/or calculated, with only the values in column A and the total column different from those in exercise 3.

	A	B	C	D	total
n	6	10	8	5	29
Σx	1479	74	147	92	1792
Σx^2	1961341	684	2951	1710	1966686
$\bar{x}$	246.500	7.400	18.375	18.400	61.793
s^2	319353.500	15.156	35.696	4.300	

$\bar{\bar{x}}$ = $\Sigma n_i\bar{x}_i/\Sigma n_i$
  = [6(246.500) + 10(7.400) + 8(18.375) + 5(18.400)]/(6 + 10 + 8 + 5)
  = 1792/29
  = 61.793   [NOTE: This must always agree with the $\bar{x}$ in the "total" column.]
$\Sigma n_i(\bar{x}_i-\bar{\bar{x}})^2$ = 6(246.500-61.793)^2 + 10(7.400-61.793)^2 + 8(18.375-61.793)^2
                  + 5(18.400-61.793)^2
      = 258781.784
$\Sigma df_i s_i^2$ = 5(31935.500) + 9(15.156) + 7(35.696) + 4(4.300)
      = 1597170.976

H_o: μ_A = μ_B = μ_C = μ_D
H_1: at least one mean is different
α = .01
C.R. F > $F^3_{25,.01}$ = 4.6755
calculations:

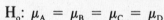

source	SS	df	MS	F
Trt	258781.784	3	86260.595	1.3502
Error	1597170.976	25	63886.839	
Total	1855952.760	28		

F = MS_{Trt}/MS_E
  = 86260/63886
  = 1.3502
conclusion:
  Do not reject H_o; there is not sufficient evidence to conclude that at least one mean is different.

The results are affected by the outlier in what may be an unexpected manner. Because the sample mean of group 1 increases dramatically, making the groups appear "more different" than before, one might expect even stronger evidence to reject H_o than the original calculated $F = 11.6743$. Instead, however, the calculated F drops to $F = 1.3502$ and the conclusion becomes that the groups are <u>not</u> significantly different at all. In general, the effect of an outlier cannot be predicted with certainty. In this case, the MS_E (the estimate of the unexplained variability [or inherent variance] within the groups) increased to the point that the differences between the sample means were no longer significant -- they were consistent with the differences expected between such small samples each having such a large variance.

NOTE: Complete the ANOVA table as follows:
(1) Enter $SS_{Trt} = \Sigma n_i(\bar{x}_i - \bar{\bar{x}})^2$ and $SS_E = \Sigma df_i s_i^2$ values from the preliminary calculations.
(2) Enter $df_{Trt} = k-1$ and $df_E = \Sigma df_i = \Sigma(n_i - 1) = \Sigma n_i - k$.
(3) Add the SS and df columns to find SS_{Tot} and df_{Tot}. [The df_{Tot} must equal $\Sigma n_i - 1$.]
(4) Calculate $MS_{Trt} = SS_{Trt}/df_{Trt}$ and $MS_E = SS_E/df_E$.
(5) Calculate $F = MS_{Trt}/MS_E$.
As a final check, calculate s^2 (i.e., the variance of all the scores in one large group) two different ways as indicated below. If these answers agree, the problem is probably correct.
    * from the "total" column in the table for the preliminary calculations:

$$s^2 = [n\Sigma x^2 - (\Sigma x)^2]/[n(n-1)]$$
$$= [29(1966686) - (1792)^2]/[29(28)]$$
$$= 53822630/812$$
$$= 66284.0$$

    * from the "total" row of the ANOVA table

$$s^2 = SS_{Tot}/df_{Tot}$$
$$= 1855952.760/28$$
$$= 66284.0$$

11-4 Two-Way ANOVA

NOTE: The formulas and principles in this section are logical extensions of the previous sections.

$SS_{Row} = \Sigma n_i(\bar{x}_i - \bar{\bar{x}})^2$ for $i=1,2,3...$ [for each row]
$SS_{Col} = \Sigma n_j(\bar{x}_j - \bar{\bar{x}})^2$ for $j=1,2,3...$ [for each column]
$SS_{Tot} = \Sigma(x - \bar{\bar{x}})^2$ [for all the x's]
When there is only one observation per cell...
    the unexplained variation is
        $SS_E = SS_{Tot} - SS_{Row} - SS_{Col}$
    and there is not enough data to measure interaction.
When there is more than one observation per cell...
    the unexplained variation (i.e., the failure of items in the same cell to respond the same) is
        $SS_E = \Sigma(x - \bar{x}_{ij})^2$ [for each cell -- i,e., for each i,j (row,col) combination]
    and the interaction sum of squares is
        $SS_{Int} = SS_{Tot} - SS_{Row} - SS_{Col} - SS_E$.

Since the data will be analyzed from statistical software packages, however, the above formulas need not be used by hand.

1. a. $MS_{Int} = 1263$
   b. $MS_E = 695$
   c. $MS_{Star} = 350$
   d. $MS_{MPAA} = 14$

3. $H_o: \mu_P = \mu_F = \mu_G = \mu_E$
   $H_1:$ at least one mean is different
   $\alpha = .05$ [or another appropriate value]
   C.R. $F > F^3_{8,.05} = 4.0662$
   calculations:
       $F = MS_{Star}/MS_E$
         $= 350/695$
         $= .5036$

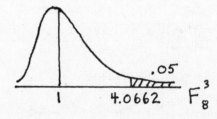

conclusion:
    Do not reject H_o; there is not sufficient evidence to conclude that at least one mean is different.
The evidence does not support a conclusion that movies of different star ratings tend to have different lengths.

5. $H_o: \mu_{G/PG/PG-13} = \mu_R$
   $H_1:$ the means are different
   $\alpha = .05$
   C.R. $F > F^1_{3,.05} = 10.128$
   calculations:
       $F = MS_{MPAA}/MS_E$
         $= 0/266$
         $= 0$
conclusion:
    Do not reject H_o; there is sufficient evidence to conclude that the means are different.

7. H_o: $\mu_A = \mu_B = \mu_C = \mu_D$
   H_1: at least one mean is different
   $\alpha = .05$
   C.R. $F > F^3_{6,.05} = 4.7571$
   calculations:
   $F = MS_{Typist}/MS_E$
   $= 7.333/.417$
   $= 17.585$
   conclusion:
   Reject H_o; there is sufficient evidence to conclude that at least one mean is different.

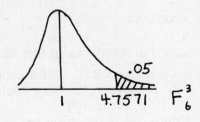

9. As indicated by the Minitab output given below, reversing the roles of the rows and columns changes the order in which the results are presented but does not change the numerical values identified with the MPAA ratings or the star ratings.

```
MTB > set c1
DATA> 108 91 98 100 93 94 103 193 105 96 110 114 115 133 72 120
DATA> end
MTB > set c2
DATA> 1 1 1 1 1 1 1 1 2 2 2 2 2 2 2 2
DATA> end
MTB > name c2'row'
MTB > set c3
DATA> 1 1 2 2 3 3 4 4 1 1 2 2 3 3 4 4
DATA> end
MTB > name c3'column'
MTB > TWOWAY C1 C2 C3

ANALYSIS OF VARIANCE   C1

SOURCE          DF        SS        MS
row              1        14        14
column           3      1049       350
INTERACTION      3      3790      1263
ERROR            8      5560       695
TOTAL           15     10413

MTB > SET C2
DATA> 1 1 2 2 3 3 4 4 1 1 2 2 3 3 4 4
DATA> END
MTB > SET C3
DATA> 1 1 1 1 1 1 1 1 2 2 2 2 2 2 2 2
DATA> END
MTB > TWOWAY C1 C2 C3

ANALYSIS OF VARIANCE   C1

SOURCE          DF        SS        MS
row              3      1049       350
column           1        14        14
INTERACTION      3      3790      1263
ERROR            8      5560       695
TOTAL           15     10413
```

11. The Minitab output at the right was obtained from the original data of exercises 1-4 and 9 as indicated. Multiplying each value in the table by 10 multiplies the entries in the df, SS and MS columns by $10^2 = 100$. [As the entries as given to the nearest whole number, there are discrepancies caused by rounding.]

```
MTB > LET C1=10*C1
MTB > TWOWAY C1 C2 C3

ANALYSIS OF VARIANCE   C1

SOURCE        DF        SS        MS
row            1      1406      1406
column         3    104919     34973
INTERACTION    3    379019    126340
ERROR          8    555950     69494
TOTAL         15   1041294
```

NOTE: All F values remain unchanged. Testing the hypothesis of exercise 3, for example,

$$F = MS_{Star}/MS_E = 34973/69494 = .5033$$

which agrees, except for a .0003 discrepancy caused by rounding, with the result given in that exercise.

Review Exercises

1. The following preliminary values are identified and/or calculated.

	A	B	C	D
n	5	5	5	5
Σx	265	295	245	325
Σx^2	14083	17415	12015	21193
$\bar{x}$	53	59	49	65
s^2	9.5	2.5	2.5	17.0

$k = 4$

$n = 5$

$s_{\bar{x}}^2 = \Sigma(\bar{x}_i - \bar{\bar{x}})^2/(k-1)$

$\quad = 147/3$

$\quad = 49$

$\bar{\bar{x}} = \Sigma\bar{x}_i/k$

$\quad = 56.5$

$s_p^2 = \Sigma s_i^2/k$

$\quad = 31.5/4$

$\quad = 7.875$

$H_o: \mu_A = \mu_B = \mu_C = \mu_D$
H_1: at least one mean is different
$\alpha = .05$
C.R. $F > F_{16,.05}^3 = 3.2389$
calculations:
$\quad F = ns_{\bar{x}}^2/s_p^2$
$\quad\quad = 5(49)/7.875$
$\quad\quad = 31.1111$
conclusion:
Reject H_o; there is sufficient evidence to conclude that at least one mean is different.

NOTE: Since $n_A = n_B = n_C = n_D$, this exercise was completed using the techniques and format of section 11-2 for equal sample sizes. It could also have been completed using the techniques and format of section 11-3 for unequal sample sizes. The method for unequal sample sizes works for all one-way ANOVA problems; the procedure given for equal sample sizes is a short-cut of that method that takes advantage of the algebraic simplification occurring when all the n's are equal.

3. The following preliminary values are identified and/or calculated.

	A	B	C	total
n	5	5	7	17
Σx	.49	.37	.33	1.19
Σx^2	.0483	.0279	.0159	.0921
$\overline{x}$	.0980	.0740	.0471	.0700
s^2	.0000700	.0001300	.0000571	

$\overline{\overline{x}} = \Sigma n_i \overline{x}_i / \Sigma n_i$
  $= [5(.0980) + 5(.0740) + 7(.0471)]/(5 + 5 + 7)$
  $= 1.19/17$
  $= .0700$ [NOTE: This must always agree with the $\overline{x}$ in the "total" column.]
$\Sigma n_i (\overline{x}_i - \overline{\overline{x}})^2 = 5(.0980-.0700)^2 + 5(.0740-.0700)^2 + 7(.0471-.0700)^2$
  $= .007657$
$\Sigma df_i s_i^2 = 4(.0000700) + 4(.0001300) + 6(.0000571)$
  $= .001143$

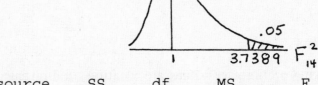

H_o: $\mu_A = \mu_B = \mu_C$
H_1: at least one mean is different
$\alpha = .05$
C.R. $F > F^2_{14,.05} = 3.7389$
calculations:

source	SS	df	MS	F
Trt	.007657	2	.003829	46.9000
Error	.001143	14	.0000816	
Total	.008800	16		

$F = MS_{Trt}/MS_E$
  $= .003829/.0000816$
  $= 46.9000$

conclusion:
  Reject H_o; there is sufficient evidence to conclude that at least one mean is different.

NOTE: Complete the ANOVA table as follows:
(1) Enter $SS_{Trt} = \Sigma n_i (\overline{x}_i - \overline{\overline{x}})^2$ and $SS_E = \Sigma df_i s_i^2$ values from the preliminary calculations.
(2) Enter $df_{Trt} = k-1$ and $df_E = \Sigma df_i = \Sigma (n_i-1) = \Sigma n_i - k$.
(3) Add the SS and df columns to find SS_{Tot} and df_{Tot}. [The df_{Tot} must equal $\Sigma n_i - 1$.]
(4) Calculate $MS_{Trt} = SS_{Trt}/df_{Trt}$ and $MS_E = SS_E/df_E$.
(5) Calculate $F = MS_{Trt}/MS_E$.
As a final check, calculate s^2 (i.e., the variance of all the scores in one large group) two different ways as indicated below. If these answers agree, the problem is probably correct.
  * from the "total" column in the table for the preliminary calculations:
    $s^2 = [n\Sigma x^2 - (\Sigma x)^2]/[n(n-1)]$
      $= [17(.0921)-(1.19)^2]/[17(16)]$
      $= .1496/272$
      $= .00055$
  * from the "total" row of the ANOVA table
    $s^2 = SS_{Tot}/df_{Tot}$
      $= .008800/16$
      $= .00055$

5. H_o: there is no engine-size/transmission-type interaction
    H_1: there is engine-size/transmission-type interaction
    $\alpha = .05$
    C.R. $F > F^2_{6,.05} = 5.1433$
    calculations:
    $\quad F = MS_{Int}/MS_E$
    $\quad\quad = .6/11.3$
    $\quad\quad = .0531$
    conclusion:

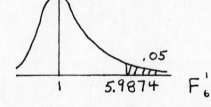

Do not reject H_o; there is not sufficient evidence to conclude that there is engine-size/transmission-type interaction.

7. H_o: $\mu_A = \mu_M$
    H_1: the means are different
    $\alpha = .05$
    C.R. $F > F^1_{6,.05} = 5.9874$
    calculations:
    $\quad F = MS_{Trans}/MS_E$
    $\quad\quad = 40.3/11.3$
    $\quad\quad = 3.5664$
    conclusion:

Do not reject H_o; there is not sufficient evidence to conclude that the means are different.

Chapter 12

Statistical Process Control

12-2 Control Charts for Variation and Mean

NOTE: In this section, n = number of observations per sample subgroup
k = number of sample subgroups

1. $\bar{\bar{x}} = \Sigma\bar{x}/k = 99.860/20 = 4.993$

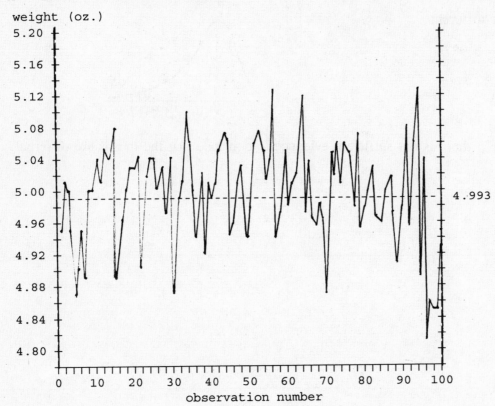

No; there does not appear to be a pattern suggesting that the process is not within statistical control.

3. $\bar{\bar{x}} = \Sigma\bar{x}/k = 99.860/20 = 4.993$
   $\bar{R} = \Sigma R/k = 2.48/20 = .124$
   $LCL = \bar{\bar{x}} - A_2\bar{R} = 1.993 - (.577)(.124) = 4.993 - .072 = 4.921$
   $UCL = \bar{\bar{x}} + A_2\bar{R} = 1.993 + (.577)(.124) = 4.993 + .072 = 5.065$

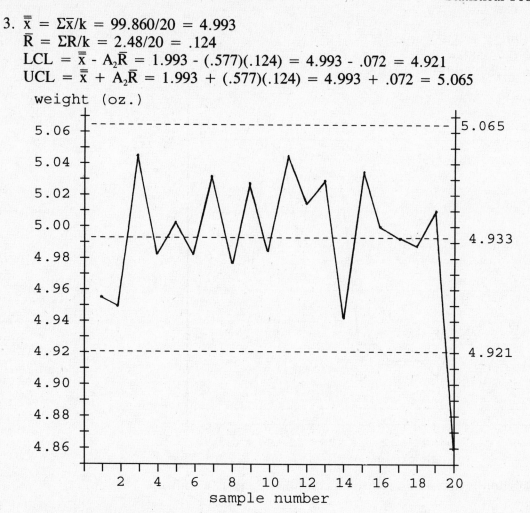

The process requires correction.

5. $\bar{R} = \Sigma R/k = 179/24 = 7.458$
   $LCL = D_3\bar{R} = 0(7.458) = 0$
   $UCL = D_4\bar{R} = 2.282(7.458) = 17.109$

time (min.)

The process variation is not within statistical control. There is a shift up.

7. $\bar{\bar{x}} = \Sigma\bar{x}/k = 20.306$

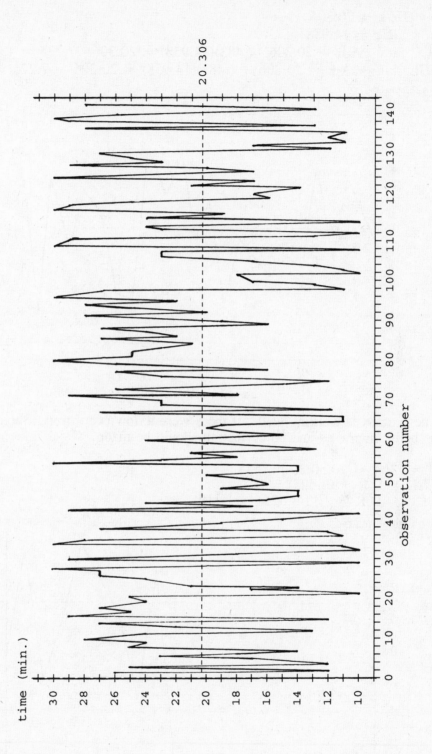

No; there does not appear to be a pattern suggesting that the process is not within statistical control.

9. $\bar{\bar{x}} = \Sigma\bar{x}/k = 20.306$

$\bar{R} = \Sigma R/k = 14.958$

$LCL = \bar{\bar{x}} - A_2\bar{R} = 20.306 - (.483)(14.958) = 20.306 - 7.225 = 13.08$

$UCL = \bar{\bar{x}} + A_2\bar{R} = 20.306 + (.483)(14.958) = 20.306 + 7.225 = 27.53$

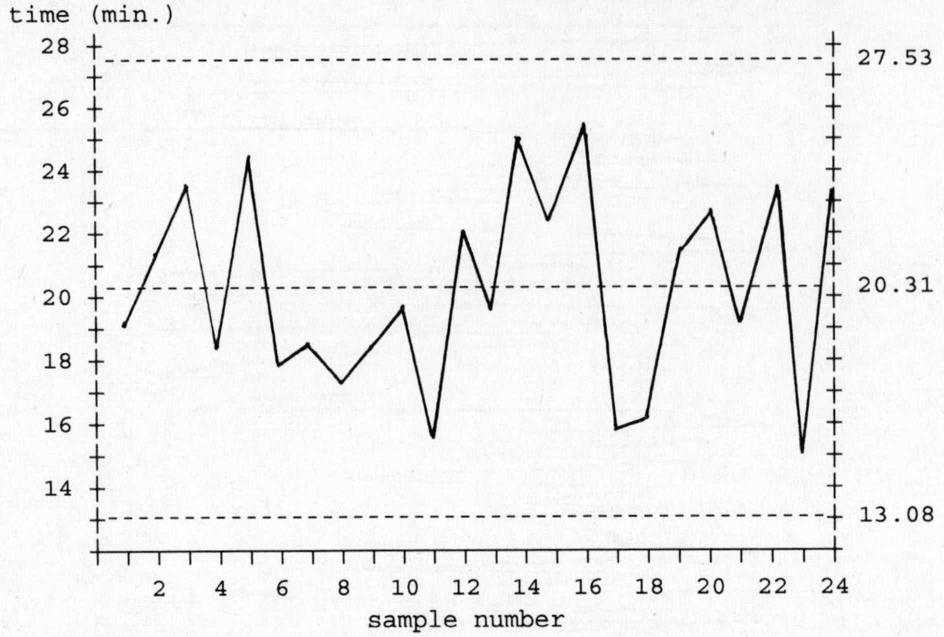

time (min.)

The process is working well. If there were a downward trend, then the delivery times would be being improved and no correction should be made.

11. $\bar{R} = \Sigma R/k = 10.50/20 = .525$

$LCL = D_3\bar{R} = 0(.525) = 0$

$UCL = D_4\bar{R} = 2.114(.525) = 1.110$

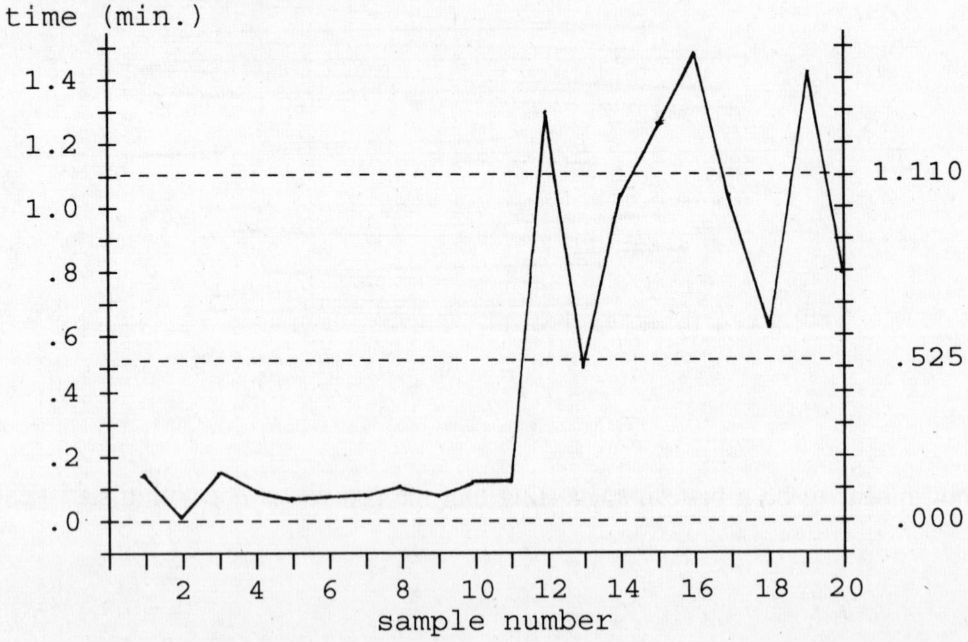

time (min.)

The process variation is not within statistical control. There is a shift up, and there are points above the upper control limit.

13. a. $\bar{s} = \Sigma s/k = 2.8979/20 = .145$
   $LCL = B_3\bar{s} = 0(.145) = 0$
   $UCL = B_4\bar{s} = 2.089(.145) = .303$

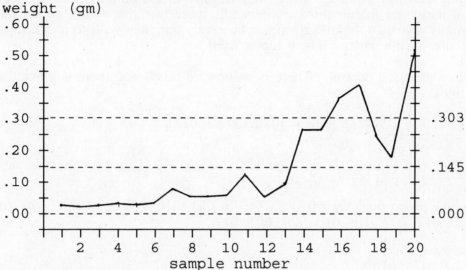

As concluded by the original method, the process is out of statistical control.

b. $\bar{\bar{x}} = \Sigma\bar{x}/k = 5.7098$ [given in the text]
   $\bar{s} = \Sigma s/k = 2.8979/20 = .145$
   $LCL = \bar{\bar{x}} - A_3\bar{s} = 5.7098 - (1.427)(.145) = 5.7098 - .2068 = 5.503$
   $UCL = \bar{\bar{x}} + A_3\bar{s} = 5.7098 + (1.427)(.145) = 5.7098 + .2068 = 5.917$

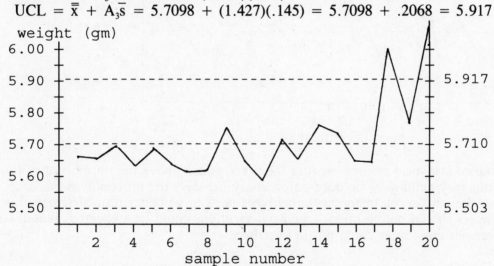

As concluded by the original method, the process is out of statistical control.

12-3 Control Charts for Attributes

1. This process is within statistical control. Since the first third of the sample means are generally less than the overall mean, the middle third are generally more than the overall mean, and the final third are generally less than the overall mean, however, one may wish to check future analyses to see whether such a patter tends to repeat itself.

3. This process is out of statistical control. There is an upward trend, and there is a point above the upper control limit.

5. $\bar{p} = (\Sigma x)/(\Sigma n) = (3+2+...+5)/(21)(300) = 108/6300 = .0171$
   $\sqrt{\bar{p}\cdot\bar{q}/n} = \sqrt{(.017)(.983)/300} = .00749$
   LCL $= \bar{p} - 3\sqrt{\bar{p}\cdot\bar{q}/n} = .0171 - 3(.00749) = .0171 - .0225 = 0$ [since it cannot be negative]
   UCL $= \bar{p} + 3\sqrt{\bar{p}\cdot\bar{q}/n} = .0171 + 3(.00749) = .0171 + .0225 = .0396$
   NOTE: The 21 sample proportions are: .010 .007 .013 .023 .010 .050 .060 .007 .020 .013 .010 .017 .013 .020 .017 .007 .013 .010 .020 .003 .017

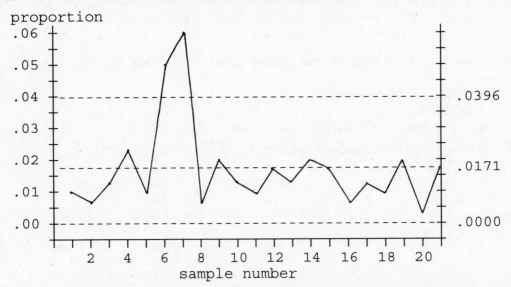

The process is out of statistical control because there are points above the upper control limit. If there were temporary employees on duty on precisely the days the process was out of statistical control, then either the temporary employees need to be better trained/screened or temporary employees should not be hired -- perhaps vacations could be staggered, output could be reduced, or regular employees could put in overtime.

7. $\bar{p} = (\Sigma x)/(\Sigma n) = (4+3+...+5)/(18)(120) = 89/2160 = .0412$

$\sqrt{\bar{p}\cdot\bar{q}/n} = \sqrt{(.041)(.959)/120} = .0181$

LCL $= \bar{p} - 3\sqrt{\bar{p}\cdot\bar{q}/n} = .0412 - 3(.0181) = .0412 - .0544 = 0$ [since it cannot be negative]

UCL $= \bar{p} + 3\sqrt{\bar{p}\cdot\bar{q}/n} = .0412 + 3(.0181) = .0412 + .0544 = .0956$

NOTE: The 18 sample proportions are: .033 .025 .025 .017 .042 .033 .025 .133 .050 .025 .025 .017 .042 .058 .067 .033 .050 .042

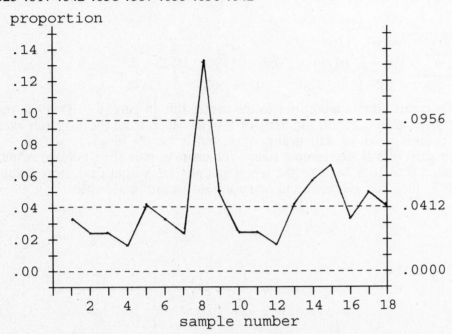

The process is out of statistical control because there is a point above the upper control limit. Since it is an isolated point, the company should examine the circumstances associated with that one specific time period (to look for a specific correctable cause -- e.g., a new person on duty) before ordering a general shutdown and/or retooling.

9. $\bar{p} = (\Sigma x)/(\Sigma n) = (10+8+...+11)/(20)(400) = 150/8000 = .01875$

a. $\hat{p} = .01875$

$\alpha = .05$

$\hat{p} \pm z_{.025}\sqrt{\hat{p}\hat{q}/n}$

$.01875 \pm 1.960\sqrt{(.01875)(.98125)/8000}$

$.01875 \pm .00297$

$.0158 < p < .0217$

b. original claim: $p \leq .01$ [normal approximation to the binomial, use z]

$H_o: p \leq .01$

$H_1: p > .01$

$\alpha = .05$

C.R. $z > z_{.05} = 1.645$

calculations:

$z_{\hat{p}} = (\hat{p} - \mu_p)/\sigma_p$

$= (.01875 - .01)/\sqrt{(.01)(.99)/8000}$

$= .00875/.00111$

$= 7.866$

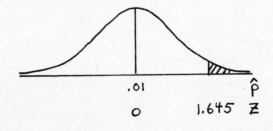

conclusion:

Reject H_o; there is sufficient evidence to conclude that $p > .01$.

11. $\bar{p} = (\Sigma x)/(\Sigma n) = .05$
In both parts (a) and (b), the center line occurs at .05.

a. n = 100
$$\sqrt{\bar{p} \cdot \bar{q}/n} = \sqrt{(.05)(.95)/100} = .0218$$
$$LCL = \bar{p} - 3\sqrt{\bar{p} \cdot \bar{q}/n} = .05 - 3(.0218) = .05 - .0654 = 0 \quad \text{[since it cannot be negative]}$$
$$UCL = \bar{p} + 3\sqrt{\bar{p} \cdot \bar{q}/n} = .05 + 3(.0218) = .05 + .0654 = .1154$$

b. n = 300
$$\sqrt{\bar{p} \cdot \bar{q}/n} = \sqrt{(.05)(.95)/300} = .0126$$
$$LCL = \bar{p} - 3\sqrt{\bar{p} \cdot \bar{q}/n} = .05 - 3(.0126) = .05 - .0378 = .0123$$
$$UCL = \bar{p} + 3\sqrt{\bar{p} \cdot \bar{q}/n} = .05 + 3(.0126) = .05 + .0378 = .0878$$

c. The lower and upper control limits are closer to the center line in part (b). This has the advantage of being better able (i.e., on the basis of less deviance from the long run average) to detect when the process is out of statistical control, but it has the disadvantage of requiring the examination of a larger sample size. The chart in part (b) would be better able to detect a shift from 5% to 10% because the larger sample size would cause less fluctuation about the 5% or 10% long run averages and make the shift more noticeable.

Review Exercises

1. $\bar{\bar{x}} = \Sigma \bar{x}/k = 5028/20 = 2.514$

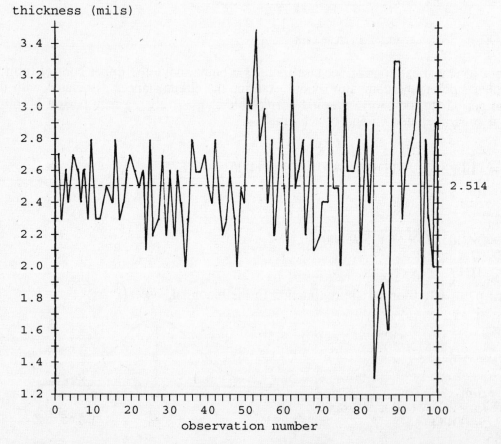

It appears that there <u>may</u> be an increase in variation.

3. $\bar{\bar{x}} = \Sigma\bar{x}/k = 5028/20 = 2.514$
   $\bar{R} = \Sigma R/k = 154/20\ .77$
   $LCL = \bar{\bar{x}} - A_2\bar{R} = 2.514 - (.577)(.77) = 2.514 - .444 = 2.070$
   $UCL = \bar{\bar{x}} + A_2\bar{R} = 2.514 + (.577)(.77) = 2.514 + .444 = 2.958$

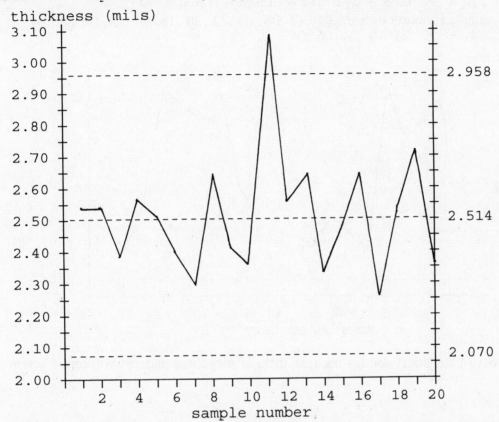

The process is out of statistical control because there is a point above the upper control limit and requires correction.

5. $\bar{p} = (\Sigma x)/(\Sigma n) = (2+6+...+4)/(24)(50) = 145/1200 = .1208$

$\sqrt{\bar{p} \cdot \bar{q}/n} = \sqrt{(.1208)(.8792)/50} = .0461$

LCL $= \bar{p} - 3\sqrt{\bar{p} \cdot \bar{q}/n} = .1208 - 3(.0461) = .1208 - .1383 = 0$ [since it cannot be negative]

UCL $= \bar{p} + 3\sqrt{\bar{p} \cdot \bar{q}/n} = .1208 + 3(.0461) = .1208 + .1383 = .2591$

NOTE: The 24 sample proportions are: .04 .12 .06 .08 .22 .20 .18 .24 .04 .06 .08 .04 .06 .10 .08 .12 .24 .20 .22 .24 .10 .04 .06 .08

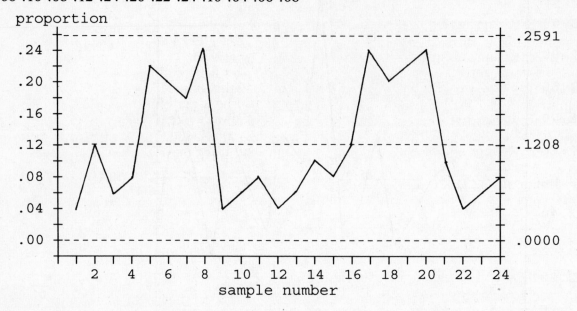

The process is out of statistical control because there is a cyclical pattern; corrective action should be taken.

Chapter 13

Nonparametric Statistics

13-2 Sign Test

NOTE: Table A-7 gives only x_L, the <u>lower</u> critical value for the sign test. Accordingly, the text lets x be the <u>smaller</u> of the number of +'s or the number of -'s and warns the user to use common sense to avoid concluding the reverse of what the data indicates.

An alternative approach maintains the natural agreement between the alternative hypothesis and the critical region and is consistent with the logic and notation of parametric tests. Let x <u>always</u> be the number of +'s. The problem's symmetry means that the upper critical value is $x_U = n\text{-}x_L$.

Since this alternative approach builds directly on established patterns and provides insight into rationale of the sign test, its C.R. and calculations and picture [notice that if H_o is true then $\mu_x = n/2$] for each exercise are given to the right of the method using only lower critical values.

1. Let Humorous be group 1.
   claim: median difference = 0

pair	A	B	C	D	E	F	G	H
H-S	+	-	+	+	+	+	+	+

   n = 8 +'s and -'s

   H_o: median difference = 0
   H_1: median difference ≠ 0
   $\alpha = .05$
   C.R. $x \le x_{L,8,.025} = 0$ <u>OR</u> C.R. $x \le x_{L,8,.025} = 0$
   $x \ge x_{U,8,.025} = 8\text{-}0 = 8$

   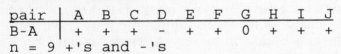

   calculations: calculations:
   x = 1   x = 7
   conclusion:
   Do not reject H_o; there is not sufficient evidence to conclude that median difference ≠ 0.

3. Let Before be group 1.
   claim: median difference = 0

pair	A	B	C	D	E	F	G	H	I	J
B-A	+	+	+	-	+	+	0	+	+	+

   n = 9 +'s and -'s

   H_o: median difference = 0
   H_1: median difference ≠ 0
   $\alpha = .05$
   C.R. $x \le x_{L,9,.025} = 1$ <u>OR</u> C.R. $x \le x_{L,9,.025} = 1$
   $x \ge x_{U,9,.025} = 9\text{-}1 = 8$

   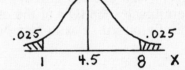

   calculations: calculations:
   x = 1   x = 8
   conclusion:
   Reject H_o; there is sufficient evidence to conclude that median difference ≠ 0 (in fact, median difference > 0).
   We can be 95% certain that the drug lowers systolic blood pressure.

5. Let Before be group 1.
   claim: median difference < 0

   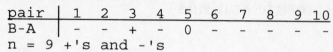

   $n = 9$ +'s and -'s

   H_o: median difference ≥ 0
   H_1: median difference < 0
   $\alpha = .05$
   C.R. $x \leq x_{L,9,.05} = 1$ OR   C.R. $x \leq x_{L,9,.05} = 1$
   calculations: calculations:
      $x = 1$ $x = 1$
   conclusion:
   Reject H_o; there is sufficient evidence to conclude that median difference < 0.
   Yes; we can be 95% certain that the new course is effective. The Dean of Curriculum should
   support the proposal.

7. Let Test I be group 1.
   claim: median difference = 0

pair	1	2	3	4	5	6	7	8	9	10
I-II	-	-	-	+	-	-	-	-	-	-

   $n = 10$ +'s and -'s

   H_o: median difference $= 0$
   H_1: median difference $\neq 0$
   $\alpha = .05$
   C.R. $x \leq x_{L,10,.025} = 1$ OR   C.R. $x \leq x_{L,10,.025} = 1$
                                          $x \geq x_{U,10,.025} = 10-1 = 9$

   calculations: calculations:
      $x = 1$ $x = 9$
   conclusion:
   Reject H_o; there is sufficient evidence to conclude that median difference $\neq 0$ (in fact,
   median difference < 0).
   We can be 95% certain that the tests give different results (and that Test I tends to give lower
   scores). Further research is necessary to determine which of the tests best measures IQ.

   NOTE: This test may not be the most appropriate one for the intended purpose. The sign test
   measures whether the overall median difference between the two tests is 0, and not whether the
   two tests give the same results for individuals. If the median difference were 0 and there were
   approximately half +'s and half -'s, the tests could still be very different. It could be, for
   example, that one test grossly overestimates the IQ's of half the students and grossly
   underestimates the IQ's of the other half -- a distinct possibility if the test is sexually or
   ethnically biased so as to help half the students and hurt the others.

9. Let Sitting be group 1.
   claim: median difference = 0

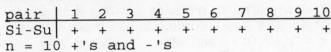

pair	1	2	3	4	5	6	7	8	9	10
Si-Su	+	+	+	+	+	+	+	+	+	+

   n = 10 +'s and -'s

   H_o: median difference = 0
   H_1: median difference $\neq$ 0
   α = .05
   C.R. x $\leq$ $x_{L,10,.025}$ = 1   OR   C.R. x $\leq$ $x_{L,10,.025}$ = 1
                                                    x $\geq$ $x_{U,10,.025}$ = 10-1 = 9

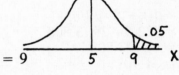

   calculations: calculations:
      x = 0   x = 10

   conclusion:
      Reject H_o; there is sufficient evidence to conclude that median difference $\neq$ 0 (in fact,
      median difference > 0).

11. Let favoring Covariant be a success.
    claim: p > .5

       7 +'s
       3 -'s
       n = 10 +'s or -'s

    H_o: p $\leq$ .5
    H_1: p > .5
    α = .05
    C.R. x $\leq$ $x_{L,10,.05}$ = 1   OR   C.R. x $\geq$ $x_{U,10,.05}$ = 10-1 = 9

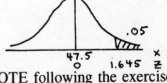

    calculations: calculations:
       x = 3   x = 7

    conclusion:
       Do not reject H_o; there is not sufficient evidence to conclude that p > .5.

13. Let the post-training scores be group 1 (i.e., let those with improved grades be +'s).
    claim: median improvement > 0

       59 +'s
       36 -'s
        5 0's
       n = 95 +'s or -'s
    Since n > 25, use z with
       μ_x = n/2 = 95/2 = 47.5
       σ_x = $\sqrt{n}$/2 = $\sqrt{95}$/2 = 4.873

    H_o: median improvement $\leq$ 0
    H_1: median improvement > 0
    α = .05
    C.R. z < $-z_{.05}$ = -1.645   OR   C.R. z > $z_{.05}$ = 1.645
    calculations: calculations: [See the NOTE following the exercise.]
       x = 36   x = 59
       z_x = $[(x+.5)-\mu_x]/\sigma_x$ z_x = $[(x-.5)-\mu_x]/\sigma_x$
          = [36.5 - 47.5]/4.873   = [58.5 - 47.5]/4.873
          = -11/4.873   = 11/4.863
          = -2.257   = 2.257
    conclusion:
       Reject H_o; there is sufficient evidence to conclude that median improvement > 0.

NOTE: The correction for continuity is a conservative adjustment intending to make less likely a false rejection of H_o by shifting the x value .5 units toward the middle. When x is the smaller of the number of +'s or the number of -'s, this always involves replacing x with x+.5. In the alternative approach, x is replaced with either x+.5 or x-.5 according to which one shifts the value toward the middle -- i.e., with x+.5 when x < μ_x = (n/2), and with x-.5 when x > μ_x = (n/2).

15. Let those greater than the hypothesized median be +'s.
claim: median = 0

```
36 +'s
22 -'s
 2 0's
 n = 58 +'s or -'s
```

Since n > 25, use z with
μ_x = n/2 = 58/2 = 29
σ_x = $\sqrt{n}$/2 = $\sqrt{58}$/2 = 3.808

H_o: median = 0
H_1: median $\neq$ 0
α = .01

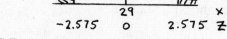

C.R. z < $-z_{.005}$ = -2.575   OR   C.R. z < $-z_{.005}$ = -2.575
                                                  z > $z_{.005}$ = 2.575

calculations: calculations:
  x = 22   x = 36
  z_x = [(x+.5)-μ_x]/σ_x z_x = [(x-.5)-μ_x]/σ_x
     = [22.5 - 29]/3.808   = [35.5 - 29]/3.808
     = -6.5/3.808   = 6.5/3.808
     = -1.707   = 1.707

conclusion:
Do not reject H_o; there is not sufficient evidence to conclude that median $\neq$ 40.

17. Let the expensive pistol scores be group 1.
claim: median difference = 0

```
24 +'s
16 -'s
 n = 40 +'s or -'s
```

Since n > 25, use z with
μ_x = n/2 = 40/2 = 20
σ_x = $\sqrt{n}$/2 = $\sqrt{40}$/2 = 3.162

H_o: median difference = 0
H_1: median difference $\neq$ 0
α = .05

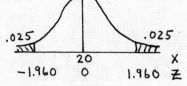

C.R. z < $-z_{.025}$ = -1.960   OR   C.R. z < $-z_{.025}$ = -1.960
                                                  z > $z_{.025}$ = 1.960

calculations: calculations:
  x = 16   x = 24
  z_x = [(x+.5)-μ_x]/σ_x z_x = [(x-.5)-μ_x]/σ_x
     = [16.5 - 20]/3.162   = [23.5 - 20]/3.162
     = -3.5/3.162   = 3.5/3.162
     = -1.107   = 1.107

conclusion:
Do not reject H_o; there is not sufficient evidence to conclude that median difference $\neq$ 0.
Yes; on the basis of this test the less expensive pistol should probably be purchased.

19. Let those greater than the hypothesized median be +'s.
    claim: median = 5.670

```
10 +'s
38 -'s
 2 0's
 n = 48 +'s or -'s
```

Since n > 25, use z with
μ_x = n/2 = 48/2 = 24
σ_x = $\sqrt{n}$/2 = $\sqrt{48}$/2 = 3.464

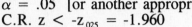

H_o: median = 5.670
H_1: median ≠ 5.670
α = .05 [or another appropriate value]
C.R. z < -$z_{.025}$ = -1.960   OR   C.R. z < -$z_{.025}$ = -1.960
                                              z > $z_{.025}$ = 1.960

calculations: calculations:
  x = 10   x = 10
  z_x = [(x+.5)-μ_x]/σ_x z_x = [(x+.5)-μ_x]/σ_x
      = [10.5 - 24]/3.464   = [10.5 - 24]/3.464
      = -13.5/3.464   = -13.5/3.464
      = -3.897   = -3.897

conclusion:
    Reject H_o; there is sufficient evidence to conclude that median ≠ 5.670 (in fact, median < 5.670).

21. The n=15 scores arranged in order are: 1 2 3 6 7 8 8 9 10 11 14 17 23 25 30.
    From Table A-7, k = 3.
    The 95% confidence interval is x_{k+1} < median < x_{n-k}
                                    x_4 < median < x_{12}
                                    6 < median < 17

NOTE: From the list of ordered scores and the fact that k = 3, it is apparent that one would reject a hypothesized median of 6 or less or of 17 or more. As in the parametric cases, the 1-α confidence interval is composed of precisely those values not rejected in a two-tailed test at the α level.

23. Let having high blood pressure be a success (i.e., those with high blood pressure are +'s).
    claim: $p \geq .5$

    ```
    n-50 +'s (the minority of the sample)
       50 -'s (the majority of the sample)
    n = # of +'s or -'s
    ```
    Since $n > 25$, use z with
       $\mu_x = n/2$
       $\sigma_x = \sqrt{n}/2$

    H_o: $p \geq .50$
    H_1: $p < .50$
    $\alpha = .01$
    C.R. $z < -z_{.01} = -2.327$
    calculations:
       $x = n - 50$

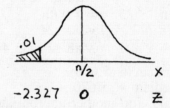

$$z_x = [(x+.5)-\mu_x]/\sigma_x \qquad \text{multiply by } \sigma_x$$
$$z_x\sigma_x = (x+.5)-\mu_x \qquad \text{substitute the given values}$$
$$-2.327 \cdot \sqrt{n}/2 = n - 50 + .5 - (n/2) \qquad \text{multiply by 2}$$
$$-2.327 \cdot \sqrt{n} = 2n - 100 + 1 - n \qquad \text{simplify}$$
$$-2.327 \cdot \sqrt{n} = n - 99 \qquad \text{square}$$
$$5.415n = n^2 - 198\,n + 9801 \qquad \text{simplify}$$
$$0 = n^2 - 203.415n + 9801 \qquad \text{use quadratic formula: } a = 1$$
$$n = [-(-203.415) \pm \sqrt{(-203.415)^2 - 4(1)(9801)}]/2(1) \qquad b = -203.415$$
$$= [203.415 \pm \sqrt{2173.633}]/2 \qquad c = 9801$$
$$= [203.415 \pm 46.662]/2$$
$$= 250.037/2 \ \text{ or } \ 156.793/2$$
$$= 125.02 \ \text{ or } \ 78.40$$

If $n \leq 78.40$, the 50 persons without high blood pressure form a large enough majority to allow the conclusion that $p < .50$. Since n must be a whole number, 78 is the largest value n can assume.
NOTE: If $n \geq 125.02$, the 50 persons without high blood pressure would form a small enough minority to allow the conclusion that $p > .50$.

13-3 Wilcoxon Signed-Ranks Test for Two Dependent Samples

NOTE: Table A-8 gives only T_L, the <u>lower</u> critical value for the signed-ranks test. Accordingly, the text lets T be the <u>smaller</u> of the sum of positive ranks or the sum of the negative ranks and warns the user to use common sense to avoid concluding the reverse of what the data indicates.

An alternative approach maintains the natural agreement between the alternative hypothesis and the critical region and is consistent with the logic and notation of parametric tests. Let T <u>always</u> be the sum of the positive ranks. By symmetry, the upper critical value is $T_U = \Sigma R - T_L$.

Since this alternative approach builds directly on established patterns and provides insight into rationale of the signed-ranks test, its C.R. and calculations and picture [notice that if H_o is true then $\mu_T = \Sigma R / 2$] for each exercise are given to the right of the method using only lower critical values.

1. Let Humorous be group 1.
    claim: the populations have the same distribution

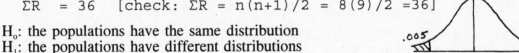

pair	A	B	C	D	E	F	G	H
H-S	2.3	-1.0	1.7	1.6	1.4	2.5	1.3	2.6
R	6	-1	5	4	3	7	2	8

    n = 8 non-zero ranks
    $\Sigma R- = 1$
    $\Sigma R+ = 35$
    $\Sigma R = 36$ [check: $\Sigma R = n(n+1)/2 = 8(9)/2 = 36$]

H_o: the populations have the same distribution
H_1: the populations have different distributions
$\alpha = .01$
C.R. $T \le T_{L,8,.005} = 0$ OR   C.R. $T \le T_{L,8,.005} = 0$
                                      $T \ge T_{U,8,.005} = 36-0 = 36$

calculations: calculations:
  $T = 1$ $T = 35$
conclusion:
    Do not reject H_o; there is not sufficient evidence to conclude that the populations have different distributions

3. Let Before be group 1.
    claim: the populations have the same distribution

pair	A	B	C	D	E	F	G	H	I	J
B-A	2	14	17	-7	17	12	0	15	33	16
R	1	4	7.5	-2	7.5	3	0	5	4	6

    n = 9 non-zero ranks
    $\Sigma R- = 2$
    $\Sigma R+ = 43$
    $\Sigma R = 45$ [check: $\Sigma R = n(n+1)/2 = 9(10)/2 = 45$]

H_o: the populations have the same distribution
H_1: the populations have different distributions
$\alpha = .05$
C.R. $T \le T_{L,9,.025} = 6$ OR   C.R. $T \le T_{L,9,.025} = 6$
                                      $T \ge T_{U,9,.025} = 45-6 = 39$

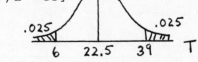

calculations: calculations:
  $T = 2$ $T = 43$
conclusion:
    Reject H_o; there is sufficient evidence to conclude that the populations have different distributions (in fact, the Before group has higher scores).

NOTE: This manual follows the text and the directions to the exercises of this section by using "the populations have the same distribution" as the null hypothesis. To be more precise, the signed-rank test doesn't test "distributions" but tests the "location" (i.e., central tendency -- as opposed to variation) of distributions. The test discerns whether one group taken as a whole tends to have higher or lower scores than another group taken as a whole. The test does not discern whether one group is more variable than another. This distinction is made clear in the wording of the conclusion when rejecting H_o in a two-tailed test (as in exercise #3 above) and when using a one-tailed test (as in exercise #5 below). For further insight into the limitations of the test and its insensitivity to differences in variation, see the NOTE accompanying exercise #7.

In addition, each exercise uses a minus sign preceding ranks associated with negative differences. While the ranks themselves are not negative, the use of the minus sign helps to organize the information.

5. Let Before be group 1.
   claim: the Before population has lower scores

pair	1	2	3	4	5	6	7	8	9	10
B-A	-36	-50	15	-2	0	-22	-36	-6	-26	-13
R	-7.5	-9	4	-1	0	-5	-7.5	-2	-6	-3

$n = 9$ non-zero ranks
$\Sigma R- = 41$
$\Sigma R+ = 4$
$\Sigma R = 45$ [check: $\Sigma R = n(n+1)/2 = 9(10)/2 = 45$]

H_o: the populations have the same distribution
H_1: the Before population has lower scores
$\alpha = .01$
C.R. $T \leq T_{L,9,.01} = 3$ <u>OR</u>   C.R. $T \leq T_{L,9,.01} = 3$
calculations: calculations:
   $T = 4$ $T = 4$

conclusion:
   Do not reject H_o; there is not sufficient evidence to conclude that the Before population has lower scores.
Since he cannot be 95% certain that the new course is effective, the Dean of Curriculum may not wish to support the proposal.

NOTE: In general, the signed-rank test of this section is a more powerful test (i.e., better able to detect departures from H_o) than the sign test of the previous section. Accordingly, one can often reject H_o with the signed-rank test but not with the sign test, or one can often find the results farther into the region of rejection with the signed-rank test than with the sign test. While it is mathematically possible to reject H_o with the sign test but not with the signed-rank test, this exercise is not a proper illustration of that phenomenon -- since H_o was rejected at the .05 level in the previous section and the .01 level was employed here. At the .05 level, the C.R. is $T \leq 6$ and H_o is rejected as it was with the sign test in the previous section.

7. Let Test I be group 1.
   claim: the populations have the same distribution

pair	1	2	3	4	5	6	7	8	9	10
I-II	-7	-9	-2	4	-4	-4	-5	-7	-7	-4
R	-8	-10	-1	3.5	-3.5	-3.5	-6	-8	-8	-3.5

   n = 10 non-zero ranks
   $\Sigma R- = 51.5$
   $\Sigma R+ = 3.5$
   $\Sigma R = 55.0$ [check: $\Sigma R = n(n+1)/2 = 10(11)/2 = 55$]

H_o: the populations have the same distribution
H_1: the populations have different distributions
$\alpha = .01$

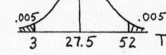

C.R. $T \le T_{L,10,.005} = 3$ OR   C.R. $T \le T_{L,10,.005} = 3$
                                        $T \ge T_{U,10,.005} = 55-3 = 52$

calculations: calculations:
  T = 3.5   T = 3.5

conclusion:
   Do not reject H_o; there is not sufficient evidence to conclude that the populations have different distributions.

NOTE: This test may not be the most appropriate one for the intended purpose. The signed-rank test measures overall differences between the groups and not whether the two tests give the same results for individuals. If one test gave half the students higher scores and the other half of the students scores that were lower by the same amounts, then $\Sigma R-$ would equal $\Sigma R+$ (so we could not reject H_o) but the distributions would be very different. This happens if one test either (a) "enhances" the scores so that the top half of the students score higher than their true IQ's and the bottom half of the students score lower than their true IQ's or (b) fails to differentiate at all so that everyone is pulled toward the middle -- then the top half of the students score lower than their true IQ's and the bottom half of the students score higher than their true IQ's.

9. Let Sitting be group 1.
   claim: the populations have the same distribution

pair	1	2	3	4	5	6	7	8	9	10
Si-Su	.99	1.60	.98	.82	1.01	1.54	.21	.70	1.67	1.32
R	5	9	4	3	6	8	1	2	10	7

   n = 10 non-zero ranks
   $\Sigma R- = 0$
   $\Sigma R+ = 55$
   $\Sigma R = 55$ [check: $\Sigma R = n(n+1)/2 = 10(11)/2 = 55$]

H_o: the populations have the same distribution
H_1: the populations have different distributions
$\alpha = .05$

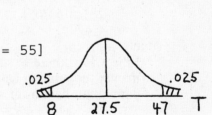

C.R. $T \le T_{L,10,.025} = 8$ OR   C.R. $T \le T_{L,10,.025} = 8$
                                        $T \ge T_{U,10,.025} = 55-8 = 47$

calculations: calculations:
  T = 0   T = 55

conclusion:
   Reject H_o; there is sufficient evidence to conclude that the populations have different distributions (in fact, the sitting positions have larger capacity scores).

11. The manual system is group 1.
    The 39 differences are listed in order at the right side of the page.

diff	rank
0	0
0	0
0	0
1	1
3	2.5
-3	-2.5
5	4.5
-5	-4.5
-6	-6
-8	-7
10	8
12	10
12	10
-12	-10
14	12.5
14	12.5
16	14.5
16	14.5
18	16
19	17.5
19	17.5
23	19
26	20
27	21.5
-27	-21.5
29	23
30	24
33	25.5
33	25.5
35	27
38	28
40	29
42	30
44	31
47	32
52	33.5
52	33.5
59	35
72	36

a. claim: the populations have the same distribution

n = 36 non-zero ranks (n > 30, use z approximation for test)

$\Sigma R- = 51.5$

$\Sigma R+ = 614.5$

$\Sigma R = 666.0$ [check: $\Sigma R = n(n+1)/2 = 36(37)/2 = 666$]

H_o: the populations have the same distribution

H_1: the populations have different distributions

$\alpha = .01$

C.R. $z < z_{.005} = -2.575$ <u>OR</u>   C.R. $z < z_{.005} = -2.575$

                                            $z > z_{.005} = 2.575$

calculations: calculations:

     T = 51.5   T = 614.5

     $z = (T - \mu_T)/\sigma_T$ $z = (T - \mu_T)/\sigma_T$

        $= (51.5 - 333)/\sqrt{36(37)(73)/24}$ $= (614.5 - 333)/63.651$

        $= -281.5/63.651$ $= 281.5/63.651$

        $= -4.423$ $= 4.423$

conclusion:

Reject H_o; there is sufficient evidence to conclude that the populations have different distributions (in fact, the manual system tends to produce longer times).

b. claim: the scanner population has lower scores

n = 36 non-zero ranks (n > 30, use z approximation for test)

$\Sigma R- = 51.5$

$\Sigma R+ = 614.5$

$\Sigma R = 666.0$ [check $\Sigma R = n(n+1)/2 = 36(37)/2 = 666$

H_o: the populations have the same distribution

H_1: the manual population has higher scores

$\alpha = .01$

C.R. $z < -z_{.01} = -2.327$ <u>OR</u>   C.R. $z > z_{.01} = 2.327$

calculations: calculations:

     T = 51.5   T = 614.5

     $z = (T - \mu_T)/\sigma_T$ $z = (T - \mu_T)/\sigma_T$

        $= (51.5 - 333)/\sqrt{36(37)(73)/24}$ $= (614.5 - 333)/63.651$

        $= -281.5/63/651$ $= 281.5/63.651$

        $= -4.423$ $= 4.423$

conclusion:

Reject H_o:; there is sufficient evidence to conclude that the manual population has higher scores -- i.e., that the scanner population has lower scores.

13. For n = 100 non-zero ranks, use the z approximation with

$\mu_T = n(n+1)/4$
$= 100(101)/4$
$= 2525$
$\sigma_T = \sqrt{n(n+1)(2n+1)/24}$
$= \sqrt{100(101)(201)/24}$
$= 290.84$

For $\alpha = .05$ in a two-tailed test, the critical z in 1.960.
The critical T (i.e., the smaller of $\Sigma R-$ and $\Sigma R+$) is found by solving as follows.

$z = (T - \mu_T)/\sigma_T < -1.96$
$(T - 2525)/290.84 < -1.96$
$T - 2525 < -570.05$
$T < 1954.95$

With the assumption of no ties in ranks, the C.R. is T ≤ 1954. NOTE: In general, however, assigning the average of the tied ranks would introduce (1/2)'s -- but not (1/3)'s, (1/4)'s, etc., since the average of n consecutive integers is always a either a whole number (if n is odd) or a half number (if n is even). The proper critical value for T would then be 1954.5.

13-4 Wilcoxon Rank-Sum Test for Two Independent Samples

1. Below are the scores (in order) for each group. The group listed first is considered group 1.

CA	R	MD	R
329	4	307	1
330	5	317	2
331	6	325	3
343	8	332	7
370	14	345	9
397	17	351	10
411	21	354	11.5
420	22	354	11.5
424	23	361	13
438	25	378	15
441	26	379	16
446	27	400	18
448	28	409	19
452	29	410	20
459	30	427	24
	285		180.0

$n_1 = 15$ $\quad \Sigma R_1 = 285$
$n_2 = 15$ $\quad \Sigma R_2 = 180$

$n = \Sigma n = 30$ $\quad \Sigma R = 465$

check: $\Sigma R = n(n+1)/2$
$= 30(31)/2$
$= 465$

$R = \Sigma R_1 = 285$

$\mu_R = n_1(n+1)/2$
$= 15(31)/2$
$= 232.5$

$\sigma_R^2 = n_1 n_2 (n+1)/12$
$= (15)(15)(31)/12$
$= 581.25$

H_o: the populations have the same distribution
H_1: the populations have different distributions
$\alpha = .05$
C.R. $z < -z_{.025} = -1.96$
$\quad\; z > z_{.025} = 1.96$
calculations:

$z_R = (R - \mu_R)/\sigma_R$
$= (285 - 232.5)/\sqrt{581.25}$
$= 52.5/24.109$
$= 2.178$

conclusion:
Reject H_o; there is sufficient evidence to conclude that the populations have different distributions (in fact, population 1 has larger salaries).
On the basis of salaries alone, California is preferred. In practice, one also must consider directly related factors (e.g., cost of living) and subjective factors (e.g., quality of living).

NOTE: As in the previous section, the manual follows the wording in the text and tests the hypothesis that "the populations have the same distribution" with the understanding that the test detects only differences in location and not differences in variability.

3. Below are the scores (in order) for each group. The group listed first is considered group 1.

O-C	R	Con	R
210	1	334	7.5
287	2	349	11
288	3	402	12
304	4	413	14
305	5	429	15
308	6	445	16
334	7.5	460	18.5
340	9	476	20.5
344	10	483	21
407	13	501	22
455	17	519	23
463	19	594	24
	96.5		203.5

$n_1 = 12$ $\Sigma R_1 = 96.5$
$n_2 = 12$ $\Sigma R_2 = 203.5$

$n = \Sigma n = 24$ $\Sigma R = 300.0$

check: $\Sigma R = n(n+1)/2$
$= 24(25)/2$
$= 300$

$R = \Sigma R_1 = 96.5$

$\mu_R = n_1(n+1)/2$
$= 12(25)/2$
$= 150$

$\sigma_R^2 = n_1 n_2 (n+1)/12$
$= (12)(12)(25)/12$
$= 300$

H_o: the populations have the same distribution
H_1: the populations have different distributions
$\alpha = .01$
C.R. $z < -z_{.005} = -2.575$
$\quad z > z_{.005} = 2.575$
calculations:

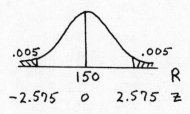

$z_R = (R - \mu_R)/\sigma_R$
$= (96.5 - 150)/\sqrt{300}$
$= -53.5/17.321$
$= -3.089$

conclusion:
Reject H_o; there is sufficient evidence to conclude that the populations have different distributions (in fact, population 1 has smaller volumes).
Based on this result, we can be 99% confident that there are biological factors related to obsessive-compulsive disorders.

5. Below are the scores (in order) for each group. The group listed first is considered group 1.

Beer	R	Liqr	R
.129	1	.182	9
.146	2	.185	10
.148	3	.190	12.5
.152	4	.205	15
.154	5	.220	17
.155	6	.224	18
.164	7	.225	19.5
.165	8	.226	20.5
.187	11	.227	21
.190	12.5	.234	22
.203	14	.241	23
.212	16	.247	24
	89.5	.253	25
		.257	26
			261.5

$n_1 = 12$ $\Sigma R_1 = 89.5$
$n_2 = 14$ $\Sigma R_2 = 261.5$

$n = \Sigma n = 26$ $\Sigma R = 351.0$

check: $\Sigma R = n(n+1)/2$
$= 26(27)/2$
$= 351$

$R = \Sigma R_1 = 89.5$

$\mu_R = n_1(n+1)/2$
$= 12(27)/2$
$= 162$

$\sigma_R^2 = n_1 n_2 (n+1)/12$
$= (12)(14)(27)/12$
$= 378$

H_o: the populations have the same distribution
H_1: the populations have different distributions
$\alpha = .05$
C.R. $z < -z_{.025} = -1.96$
      $z > z_{.025} = 1.96$
calculations:
      $z_R = (R - \mu_R)/\sigma_R$
      $= (89.5 - 162)/\sqrt{378}$
      $= -72.5/19.442$
      $= -3.729$

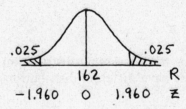

conclusion:
      Reject H_o; there is sufficient evidence to conclude that the populations have different distributions (in fact, population 1 has lower BAC levels).
Based on these results, it appears that the liquor drinkers have a higher BAC level and are (presumably) more dangerous.

7. Below are the scores (in order) for each group. The group listed first is considered group 1.

Exp	R	Cont	R
21.37	1	54.22	5
34.26	2	70.02	13
44.09	3	70.70	14
44.71	4	74.01	18
57.29	6	94.23	24
59.77	7	99.08	25
59.78	8	104.06	26
60.83	9	111.26	27
64.05	10	118.43	29
66.87	11	118.58	30
69.95	12	119.89	31
72.14	15	120.76	32
73.46	16	121.67	33
74.27	17	121.70	34
75.38	19	122.80	35
76.59	20	138.27	36
80.03	21		412
82.25	22		
92.72	23		
117.80	28		
	254		

$n_1 = 20 \qquad \Sigma R_1 = 254$
$n_2 = 16 \qquad \Sigma R_2 = 412$

$n = \Sigma n = 36 \qquad \Sigma R = 666$

check: $\Sigma R = n(n+1)/2$
$= 36(37)/2$
$= 666$

$R = \Sigma R_1 = 254$

$\mu_R = n_1(n+1)/2$
$\quad = 20(37)/2$
$\quad = 370$

$\sigma_R^2 = n_1 n_2 (n+1)/12$
$\quad = (20)(16)(37)/12$
$\quad = 986.67$

H_0: the populations have the same distribution
H_1: the populations have different distributions
$\alpha = .05$
C.R. $z < -z_{.025} = -1.96$
$\quad\ \ z > z_{.025} = 1.96$
calculations:
$z_R = (R - \mu_R)/\sigma_R$
$\quad = (254 - 370)/\sqrt{986.67}$
$\quad = -116/31.411$
$\quad = -3.693$

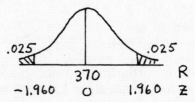

conclusion:
Reject H_0; there is sufficient evidence to conclude that the populations have different distributions (in fact, population 1 has lower scores).

9. Below are the scores (in order) for each group. The group listed first is considered group 1.

Male	R	Female	R
12.27	2	6.66	1
12.63	3	13.88	4
17.98	8	13.96	5
18.06	9	14.34	6
19.54	12	15.89	7
19.84	13	18.71	10
20.20	15	19.20	11
22.12	16	20.15	14
22.99	17	23.90	19
23.01	18	29.85	23
23.93	20	31.13	24
25.63	21		124
25.73	22		
32.20	25		
32.56	26		
39.53	27		
	254		

$n_1 = 16 \qquad \Sigma R_1 = 254$
$n_2 = 11 \qquad \Sigma R_2 = 124$
———————— ————————
$n = \Sigma n = 27 \qquad \Sigma R = 378$

check: $\Sigma R = n(n+1)/2$
$\qquad\qquad = 27(28)/2$
$\qquad\qquad = 378$

$R = \Sigma R_1 = 254$

$\mu_R = n_1(n+1)/2$
$\qquad = 16(28)/2$
$\qquad = 224$

$\sigma_R^2 = n_1 n_2 (n+1)/12$
$\qquad = (16)(11)(28)/12$
$\qquad = 410.67$

H_o: the populations have the same distribution
H_1: the populations have different distributions
$\alpha = .05$
C.R. $z < -z_{.025} = -1.96$
$\qquad z > z_{.025} = 1.96$
calculations:
$z_R = (R - \mu_R)/\sigma_R$
$\qquad = (254 - 224)/\sqrt{410.67}$
$\qquad = 30/20.265$
$\qquad = 1.480$

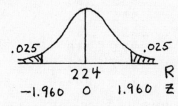

conclusion:
Do not reject H_o; there is not sufficient evidence to conclude that the populations have different distributions.

11. Below are the scores (in order) for each group. The group listed first is considered group 1.

red	R	yel	R
.817	1	.822	2
.844	3	.845	4
.864	6	.861	5
.870	7	.890	10
.871	8	.890	11
.876	9	.892	12
.896	13	.905	14
.909	15	.912	17
.913	19	.912	17
.920	22.5	.912	17
.921	24	.914	20.5
.931	30.5	.914	20.5
.933	32	.920	22.5
.941	35	.922	25
.952	36	.925	26
.957	37	.927	27
.958	38.5	.929	28
	336.5	.930	29
		.931	30.5
		.935	33
		.936	34
		.958	38.5
		.971	40
		.986	41
			524.5

$$n_1 = 17 \qquad \Sigma R_1 = 336.5$$
$$n_2 = 24 \qquad \Sigma R_2 = 524.5$$

$$n = \Sigma n = 41 \qquad \Sigma R = 861.0$$

$$\text{check: } \Sigma R = n(n+1)/2$$
$$= 41(42)/2$$
$$= 861$$

$$R = \Sigma R_1 = 336.5$$

$$\mu_R = n_1(n+1)/2$$
$$= 17(42)/2$$
$$= 357$$

$$\sigma_R^2 = n_1 n_2 (n+1)/12$$
$$= (17)(24)(42)/12$$
$$= 1428$$

H_o: the populations have the same distribution
H_1: the populations have different distributions
$\alpha = .05$ [or another appropriate value]
C.R. $z < -z_{.025} = -1.96$
$\qquad z > z_{.025} = 1.96$
calculations:
$$z_R = (R - \mu_R)/\sigma_R$$
$$= (336.5 - 357)/\sqrt{1428}$$
$$= -20.5/37.789$$
$$= -.542$$
conclusion:
Do not reject H_o; there is not sufficient evidence to conclude that the populations have different distributions.

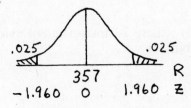

13. a. The format of the previous exercises produces the following. Consider the A's group 1.

A	R	B	R
	1		16
	2		17
	3		18
	4		19
	5		20
	6		21
	7		22
	8		23
	9		24
	10		25
	11		26
	12		27
	13		28
	14		29
	15		30
	120		345

$n_1 = 15$ $\Sigma R_1 = 120$
$n_2 = 15$ $\Sigma R_2 = 345$

$n = \Sigma n = 30$ $\Sigma R = 465$

check: $\Sigma R = n(n+1)/2$
$\qquad\quad = 30(31)/2$
$\qquad\quad = 465$

$R = \Sigma R_1 = 120$

$\mu_R = n_1(n+1)/2$
$\quad\ = 15(31)/2$
$\quad\ = 232.5$

$\sigma_R^2 = n_1 n_2(n+1)/12$
$\quad\ = (15)(15)(31)/12$
$\quad\ = 581.25$

H_o: the populations have the same distribution
H_1: the populations have different distributions
$\alpha = .05$
C.R. $z < -z_{.025} = -1.96$
$\qquad z > z_{.025} = 1.96$
calculations:

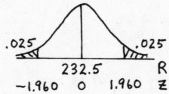

$z_R = (R - \mu_R)/\sigma_R$
$\quad\ = (120 - 232.5)/\sqrt{581.25}$
$\quad\ = -112.5/24.109$
$\quad\ = -4.666$

conclusion:

Reject H_o; there is sufficient evidence to conclude that the populations have different distributions (in fact, population A has lower scores).

NOTE: These are the most one-sided results possible for population A having lower scores than population B -- every score from A was lower than the lowest score from B. The statistical conclusion (i.e., rejecting H_o) is the one demanded by common sense. With these sample sizes it is not possible to have a calculated z less than -4.666.

b. The format of the previous exercises produces the following. Consider the A's group 1.

A	R	B	R
	1		2
	3		4
	5		6
	7		8
	9		10
	11		12
	13		14
	15		16
	17		18
	19		20
	21		22
	23		24
	25		26
	27		28
	29		30
	225		240

$n_1 = 15$ $\Sigma R_1 = 225$
$n_2 = 15$ $\Sigma R_2 = 240$
_____
$n = \Sigma n = 30$ $\Sigma R = 465$

check: $\Sigma R = n(n+1)/2$
                $= 30(31)/2$
                $= 465$

$R = \Sigma R_1 = 225$

$\mu_R = n_1(n+1)/2$
     $= 15(31)/2$
     $= 232.5$

$\sigma_R^2 = n_1 n_2 (n+1)/12$
      $= (15)(15)(31)/12$
      $= 581.25$

H_o: the populations have the same distribution
H_1: the populations have different distributions
$\alpha = .05$
C.R. $z < -z_{.025} = -1.96$
     $z > z_{.025} = 1.96$
calculations:

$z_R = (R - \mu_R)/\sigma_R$
   $= (225 - 232.5)/\sqrt{581.25}$
   $= -7.5/24.109$
   $= -.311$

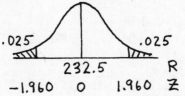

conclusion:
   Do not reject H_o; there is not sufficient evidence to conclude that the populations have different distributions.

NOTE: These results represent an alternating inter-mixing of A scores and B scores. The statistical conclusion (i.e., not rejecting H_o) is the one demanded by common sense.

c. Parts (a) and (b) represent two possible extreme data patterns. In part (a), all the low scores in one group and all the high scores in the other; in part (b), the ordered scores alternate between the two groups. Common sense demands that the hypothesis of identical population distributions be rejected in part (a) and not be rejected in part (b). The statistical conclusions support the demands of common sense.

d. The format of the previous exercises produces the following. Consider the A's group 1.

A	R	B	R
	16		1
	17		2
	18		3
	19		4
	20		5
	21		6
	22		7
	23		8
	24		9
	25		10
	26		11
	27		12
	28		13
	29		14
	30		15
	345		120

$$n_1 = 15 \qquad \Sigma R_1 = 345$$
$$n_2 = 15 \qquad \Sigma R_2 = 120$$

$$n = \Sigma n = 30 \qquad \Sigma R = 465$$

check: $\Sigma R = n(n+1)/2$
$$= 30(31)/2$$
$$= 465$$

$$R = \Sigma R_1 = 345$$

$$\mu_R = n_1(n+1)/2$$
$$= 15(31)/2$$
$$= 232.5$$

$$\sigma_R^2 = n_1 n_2 (n+1)/12$$
$$= (15)(15)(31)/12$$
$$= 581.25$$

H_o: the populations have the same distribution
H_1: the populations have different distributions
$\alpha = .05$
C.R. $z < -z_{.025} = -1.96$
$\qquad z > z_{.025} = 1.96$
calculations:

$$z_R = (R - \mu_R)/\sigma_R$$
$$= (345 - 232.5)/\sqrt{581.25}$$
$$= 112.5/24.109$$
$$= 4.666$$

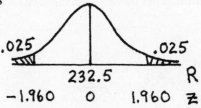

conclusion:

Reject H_o; there is sufficient evidence to conclude that the populations have different distributions (in fact, population A has higher scores).

Compared to part (a), the R value changes from 120 to 345 -- the μ_R and σ_R, which depend only on the sample sizes, remain the same. This has the effect of reversing the sign on the calculated z statistic.

NOTE: Part (c) is the "mirror image" of part (a) -- these are the most one-sided results possible for population A having higher scores than population B -- every score from A was higher than the highest score from B. The statistical conclusion (i.e., rejecting H_o) is the one demanded by common sense. With these sample sizes it is not possible to have a calculated z greater than 4.666.

e. The format of the previous exercises produces the following. Consider the A's group 1.

A R	B R		
30	16	$n_1 = 15$	$\Sigma R_1 = 149$
2	17	$n_2 = 15$	$\Sigma R_2 = 316$
3	18		
4	19	$n = \Sigma n = 30$	$\Sigma R = 465$
5	20		
6	21	check: $\Sigma R = n(n+1)/2$	
7	22	$= 30(31)/2$	
8	23	$= 465$	
9	24		
10	25	$R = \Sigma R_1 = 149$	
11	26		
12	27	$\mu_R = n_1(n+1)/2$	
13	28	$= 15(31)/2$	
14	29	$= 232.5$	
15	1	$\sigma_R^2 = n_1 n_2 (n+1)/12$	
149	316	$= (15)(15)(31)/12$	
		$= 581.25$	

H_o: the populations have the same distribution
H_1: the populations have different distributions
$\alpha = .05$
C.R. $z < -z_{.025} = -1.96$
$\quad\quad z > z_{.025} = 1.96$
calculations:

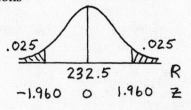

$\quad z_R = (R - \mu_R)/\sigma_R$
$\quad\quad = (149 - 232.5)/\sqrt{581.25}$
$\quad\quad = -83.5/24.109$
$\quad\quad = -3.463$
conclusion:

Reject H_o; there is sufficient evidence to conclude that the populations have different distributions (in fact, population A has lower scores).

Compared to part (a), the R value changes from 120 to 149 -- the μ_R and σ_R, which depend only on the sample sizes, remain the same. This has the effect of moving the calculated z statistic away from its negative limit of -4.667 to a less dramatic (but still significant) -3.463.

NOTE: These are strange results. Every score from A except one was lower than the second lowest score from B -- and that one exception from A was the highest score of all, while the lowest score from B was the lowest score of all. Common sense suggests that the populations from which the samples come are highly skewed -- that the A scores are basically low with a very small percentage of high scores, and that the B scores are basically high with a very small percentage of low scores. This being the case, the scores from A are generally lower than the scores from B and we hope the statistical test would so conclude -- as indeed it did..

15. With $n_1 = 2$ and $n_2 = 2$ (so that $n = n_1 + n_2 = 4$), the test of hypothesis is the same as in the previous exercises except that the samples are not large enough to use the normal approximation and the critical region must be based on $R = \Sigma R_1$ instead of z. Let the A's be considered group 1.

H_o: the populations have the same distribution
H_1: the populations have different distributions
$\alpha = .05$
C.R. R < ?
     R > ?

a. There are 6 possible arrangements of 2 scores from A and 2 scores from B. Those 6 arrangements and their associated $R = \Sigma R_1$ values are as follows.

```
rank
1 2 3 4 | R
A A B B | 3
A B A B | 4
A B B A | 5
B A A B | 5
B A B A | 6
B B A A | 7
```

b. If H_o is true and populations A and B are identical, then there is no reason to expect any of the arrangements over another and each of the 6 arrangements is equally likely. The possible values of R and the probability associated with value is therefore as given below.

R	P(R)	NOTE: R·P(R)	R^2·P(R)
3	1/6	3/6	9/6
4	1/6	4/6	16/6
5	2/6	10/6	50/6
6	1/6	6/6	36/6
7	1/6	7/6	49/6
	1	30/6	160/6

$\mu_R = \Sigma R \cdot P(R)$
$= 30/6 = 5$
$\sigma_R^2 = \Sigma R^2 \cdot P(R) - (\mu_R)^2$
$= 160/6 - 5^2$
$= 10/6$

NOTE: The fact that each of the arrangements is equally likely can be proven using conditional probability formulas for selecting A's and B's from a finite population of 2 A's and 2 B's as follows.

$P(A_1 A_2 B_3 B_4) = P(A_1) \cdot P(A_2) \cdot P(B_3) \cdot P(B_4) = (2/4) \cdot (1/3) \cdot (2/2) \cdot (1/1) = 4/24 = 1/6$
$P(A_1 B_2 A_3 B_4) = P(A_1) \cdot P(B_2) \cdot P(A_3) \cdot P(B_4) = (2/4) \cdot (2/3) \cdot (1/2) \cdot (1/1) = 4/24 = 1/6$
$P(A_1 B_2 B_3 A_4) = P(A_1) \cdot P(B_2) \cdot P(B_3) \cdot P(A_4) = (2/4) \cdot (2/3) \cdot (1/2) \cdot (1/1) = 4/24 = 1/6$
$P(B_1 A_2 A_3 B_4) = P(B_1) \cdot P(A_2) \cdot P(A_3) \cdot P(B_4) = (2/4) \cdot (2/3) \cdot (1/2) \cdot (1/1) = 4/24 = 1/6$
$P(B_1 A_2 B_3 A_4) = P(B_1) \cdot P(A_2) \cdot P(B_3) \cdot P(A_4) = (2/4) \cdot (2/3) \cdot (1/2) \cdot (1/1) = 4/24 = 1/6$
$P(B_1 B_2 A_3 A_4) = P(B_1) \cdot P(B_2) \cdot P(A_3) \cdot P(A_4) = (2/4) \cdot (1/3) \cdot (2/2) \cdot (1/1) = 4/24 = 1/6$

In addition, the μ_R and σ_R^2 values calculated above using the probability formulas agree with the formulas
$\mu_R = n_1(n+1)/2 = 2(5)/2 = 5$
$\sigma_R^2 = n_1 n_2(n+1)/12 = (2)(2)(5)/12 = 20/12 = 10/6$
given in this section. Since the distribution -- as evidenced by the P(R) values -- is not normal, however, we cannot use these values to convert to z scores for use with Table A-2.

c. The P(R) values indicate that the smallest α at which the test can be performed is the $2/6 = .333$ level with
    C.R. R ≤ 3
        R ≥ 7
as illustrated at the right -- using discrete bars instead of a continuous normal distribution. Even with the smallest or largest possible R values (i.e., R = 3 or R = 7) it would not be possible to reject H_o using $\alpha = .05$.

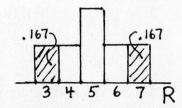

13-5 Kruskal-Wallis Test

NOTE: As in the previous sections, the manual follows the wording in the text and tests the hypothesis that "the populations have the same distribution" with the understanding that the test detects only differences in location and not differences in variability.

1. Below are the scores (in order) for each group. The group listed first is group 1, etc.

18-20	R	21-29	R	30+	R
97.1	2	97.9	6	97.4	1
97.7	5	98.2	8.5	97.5	3.5
98.0	7	98.2	8.5	97.5	3.5
98.4	10	99.0	14	98.6	12.5
98.5	11	99.6	15	98.6	12.5
	35		52.0		33.0

$n_1 = 5$ $R_1 = 35$
$n_2 = 5$ $R_2 = 52$
$n_3 = 5$ $R_3 = 33$
_____
$n = \Sigma n = 15$ $\Sigma R = 120$

check:
$\Sigma R = n(n+1)/2$
$= 15(16)/2$
$= 120$

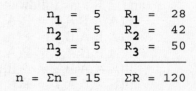

H_o: the populations have the same distribution
H_1: the populations have different distributions
$\alpha = .05$
C.R. $H > \chi^2_{2,.05} = 5.991$
calculations:
$$H = [12/n(n+1)] \cdot [\Sigma(R_i^2/n_i)] - 3(n+1)$$
$$= [12/15(16)] \cdot [(35)^2/5 + (52)^2/5 + (33)^2/5] - 3(16)$$
$$= [.05] \cdot [1003.6] - 48$$
$$= 2.180$$
conclusion:
   Do not reject H_o; there is not sufficient evidence to conclude that the populations have different distributions.

3. Below are the scores (in order) for each group. The group listed first is group 1, etc.

T	R	L	R	E	R
25.9	1	27.0	4	26.3	3
26.0	2	27.6	7	27.1	5
27.3	6	28.1	8	29.2	13
28.2	9	28.8	11	29.8	14
28.5	10	29.0	12	30.0	15
	28		42		50

$n_1 = 5$ $R_1 = 28$
$n_2 = 5$ $R_2 = 42$
$n_3 = 5$ $R_3 = 50$
_____
$n = \Sigma n = 15$ $\Sigma R = 120$

check:
$\Sigma R = n(n+1)/2$
$= 15(16)/2$
$= 120$

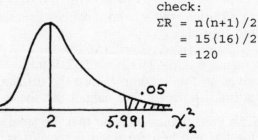

H_o: the populations have the same distribution
H_1: the populations have different distributions
$\alpha = .05$
C.R. $H > \chi^2_{2,.05} = 5.991$
calculations:
$$H = [12/n(n+1)] \cdot [\Sigma(R_i^2/n_i)] - 3(n+1)$$
$$= [12/15(16)] \cdot [(28)^2/5 + (42)^2/5 + (50)^2/5] - 3(16)$$
$$= [.05] \cdot [1009.6] - 48$$
$$= 2.480$$
conclusion:
   Do not reject H_o; there is not sufficient evidence to conclude that the populations have different distributions.

5. Below are the scores (in order) for each group. The group listed first is group 1, etc.

1	R	2	R	3	R	4	R	5	R
2.9	9.5	2.7	2	2.8	5.5	2.7	2	2.8	5.5
2.9	9.5	3.2	19.5	2.8	5.5	2.7	2	3.1	15
3.0	12	3.3	24	2.8	5.5	2.9	9.5	3.5	31.5
3.1	15	3.4	28	3.2	19.5	2.9	9.5	3.7	36
3.1	15	3.4	28	3.3	24	3.2	19.5	4.1	44
3.1	15	3.6	34	3.3	24	3.2	19.5	4.1	44
3.1	15	3.8	39	3.5	31.5	3.3	24	4.2	46.5
3.7	36	3.8	39	3.5	31.5	3.3	24		222.5
4.7	36	4.0	42	3.5	31.5	3.4	28		
3.9	41	4.1	44	3.8	39		138.0		
4.2	46.5	4.3	48		217.5				
	250.5		347.5						

$n_1 = 11$ $R_1 = 250.5$
$n_2 = 11$ $R_2 = 347.5$
$n_3 = 10$ $R_3 = 217.5$
$n_4 = 9$ $R_4 = 138$
$n_5 = 7$ $R_5 = 222.5$

$n = \Sigma n = 48$ $\Sigma R = 1176.0$

check:
$\Sigma R = n(n+1)/2$
$= 48(49)/2$
$= 1176$

H_o: the populations have the same distribution
H_1: the populations have different distributions
$\alpha = .10$
C.R. $H > \chi^2_{4,.10} = 7.779$
calculations:

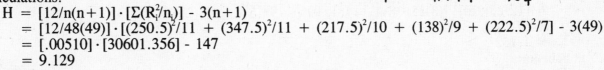

$H = [12/n(n+1)] \cdot [\Sigma(R_i^2/n_i)] - 3(n+1)$
$= [12/48(49)] \cdot [(250.5)^2/11 + (347.5)^2/11 + (217.5)^2/10 + (138)^2/9 + (222.5)^2/7] - 3(49)$
$= [.00510] \cdot [30601.356] - 147$
$= 9.129$

conclusion:
Reject H_o; there is sufficient evidence to conclude that the populations have different distributions.

7. Below are the scores for each group. The group listed first is group 1, etc.

1	R	4	R	7	R
20	19	18	14.5	21	21
18	14.5	20	19	16	9
27	24	19	17	14	6
17	11	12	3.5	23	22
18	14.5	9	1	17	11
13	5	18	14.5	17	11
24	23	28	25	20	19
15	7.5	10	2		99
	118.5	15	7.5		
		12	3.5		
			107.5		

$n_1 = 8$ $R_1 = 118.5$
$n_2 = 10$ $R_2 = 107.5$
$n_3 = 7$ $R_3 = 99.0$

$n = \Sigma n = 25$ $\Sigma R = 325.0$

check:
$\Sigma R = n(n+1)/2$
$= 25(26)/2$
$= 325$

H_o: the populations have the same distribution
H_1: the populations have different distributions
$\alpha = .05$
C.R. $H > \chi^2_{2,.05} = 5.991$
calculations:
$H = [12/n(n+1)] \cdot [\Sigma(R_i^2/n_i)] - 3(n+1)$
$= [12/25(26)] \cdot [(118.5)^2/8 + (107.5)^2/10 + (99)^2/7] - 3(26)$
$= [.0185] \cdot [4311.049] - 78$
$= 1.589$

conclusion:
Do not reject H_o; there is not sufficient evidence to conclude that the populations have different distributions.

9. Below are the scores for each group. The group listed first is group 1, etc.

1	R	4	R	7	R
1.9	13.5	2.8	23	2.7	22
2.4	21	1.8	11.5	1.9	13.5
1.5	3.5	3.2	24	1.3	1
1.6	6	2.1	16.5	1.7	9
1.6	6	1.4	2	2.2	19
1.5	3.5	2.1	16.5	2.0	15
1.7	9	4.2	25	2.2	19
2.2	19	1.7	9		98.5
	81.5	1.8	11.5		
		1.6	6		
			145.0		

$n_1 = 8$ $R_1 = 81.5$
$n_2 = 10$ $R_2 = 145.0$
$n_3 = 7$ $R_3 = 98.5$

$n = \Sigma n = 25$ $\Sigma R = 325.0$

check:
$\Sigma R = n(n+1)/2$
$= 24(25)/2$
$= 325$

H_o: the populations have the same distribution
H_1: the populations have different distributions
$\alpha = .05$
C.R. $H > \chi^2_{2,.05} = 5.991$
calculations:

$H = [12/n(n+1)] \cdot [\Sigma(R_i^2/n_i)] - 3(n+1)$
$= [12/25(26)] \cdot [(81.5)^2/8 + (145)^2/10 + (98.5)^2/7] - 3(26)$
$= [.0217] \cdot [3381.6] - 78$
$= 1.732$

conclusion:

Do not reject H_o; there is not sufficient evidence to conclude that the populations have different distributions.

11. Below are the scores (in order) for each group. The group listed first is group 1, etc.

below 2750	R	2750- 2999	R	3000- 3499	R	above 3500	R
27	9	24	2	27	9	21	1
29	20	26	5	28	15	25	3
30	23.5	27	9	28	15	26	5
30	23.5	28	15	28	15	26	5
31	26.5	28	15	28	15	27	9
33	29	28	15	29	20	27	9
34	30	28	15	30	23.5		32
35	31	29	20	30	23.5		
32	32	31	26.5		121.0		
	224.5	32	28				
			150.5				

$n_1 = 9$ $R_1 = 224.5$
$n_2 = 10$ $R_2 = 150.5$
$n_3 = 7$ $R_3 = 121$
$n_4 = 6$ $R_4 = 32$

$n = \Sigma n = 32$ $\Sigma R = 528.0$

check:
$\Sigma R = n(n+1)/2$
$= 32(33)/2$
$= 528$

H_o: the populations have the same distribution
H_1: the populations have different distributions
$\alpha = .05$
C.R. $H > \chi^2_{3,.05} = 7.815$
calculations:

$H = [12/n(n+1)] \cdot [\Sigma(R_i^2/n_i)] - 3(n+1)$
$= [12/32(33)] \cdot [(224.5)^2/9 + (150.5)^2/10 + (121)^2/7 + (32)^2/6] - 3(33)$
$= [.01136] \cdot [10127.29] - 99$
$= 16.083$

conclusion:

Reject H_o; there is sufficient evidence to conclude that the populations have different distributions.

13. a. $H = [12/n(n+1)] \cdot [\Sigma(R_i^2/n_i)] - 3(n+1)$ where $n = \Sigma n_i = 8(6) = 48$

$= [12/48(49)] \cdot [\Sigma(R_i^2/6)] - 3(49)$

$= [1/196] \cdot (1/6)[\Sigma(R_i^2)] - 147$

$= [1/1176] \cdot [R_1^2 + R_2^2 + R_3^2 + R_4^2 + R_5^2 + R_6^2 + R_7^2 + R_8^2] - 147$

b. Since adding or subtracting a constant to each score does not affect the order of the scores, their ranks and the calculated H statistic are not affected.

c. Since multiplying or dividing each score by a positive constant does not affect the order of the scores, their ranks and the calculated H statistic are not affected.

15. NOTE: Be careful when counting the number of tied ranks; in addition to the easily recognized ".5's," there are 3 12's, 3 16's and 3 23's. The following table organizes the calculations.

rank	t	$T = t^3 - t$
4.5	2	6
7.5	2	6
9.5	2	6
12	3	24
16	3	24
19.5	2	6
23	3	24
29.5	4	60
32.5	2	6
34.5	2	6
37.5	2	6
40.5	2	6
48.5	2	6
		186

correction factor:

$1 - \Sigma T/(n^3-n) = 1 - 186/(60^3-60)$

$= 1 - 186/215940$

$= 1 - .000861$

$= .999139$

The original calculated test statistic is $H = 3.489$.
The corrected calculated test statistic is $H = 3.489/.999139 = 3.492$.

13-6 Rank Correlation

1. a. Since $n \leq 30$, use Table A-9. CV: $r_s = \pm.450$
   b. Since $n > 30$, use Formula 13-1. CV: $r_s = \pm1.960/\sqrt{49} = \pm.280$
   c. Since $n > 30$, use Formula 13-1. CV: $r_s = \pm2.327/\sqrt{39} = \pm.373$

NOTE: This manual calculates $d = R_x - R_y$, thus preserving the sign of d. This convention means Σd must equal 0 and provides a check for the assigning and differencing of the ranks.

3. The following table summarizes the calculations.

x	R_x	y	R_y	d	d^2
63	2	43	5	-3	9
68	3	44	6	-3	9
71	5	39	4	1	1
55	1	30	3	-2	4
70	4	28	2	2	4
75	6	20	1	5	25
				0	52

$r_s = 1 - [6(\Sigma d^2)]/[n(n^2-1)]$

$= 1 - [6(52)]/[6(35)]$

$= 1 - 1.486$

$= -.486$

5. The following table summarizes the calculations.

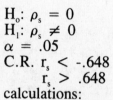

R_x	R_y	d	d^2
2	5	-3	9
6	2	4	16
3	3	0	0
3	8	-3	9
7	10	-3	9
10	9	1	1
9	1	8	64
8	7	1	1
4	6	-2	4
1	4	-3	9
		0	122

$$r_s = 1 - [6(\Sigma d^2)] / [n(n^2-1)]$$
$$= 1 - [6(122)] / [10(99)]$$
$$= 1 - .739$$
$$= .261$$

$H_o: \rho_s = 0$
$H_1: \rho_s \neq 0$
$\alpha = .05$
C.R. $r_s < -.648$
$\quad\quad r_s > .648$
calculations:
$\quad r_s = .261$
conclusion:
Do not reject H_o; there is not sufficient evidence to conclude that $\rho_s \neq 0$.

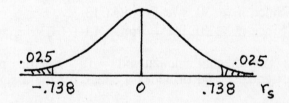

7. The following table summarizes the calculations.

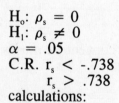

R_x	R_y	d	d^2
1	4	-3	9
2	3	-1	1
7	5	2	4
4	6	-2	4
6	7	-1	1
8	8	0	0
3	2	1	1
5	1	4	16
		0	36

$$r_s = 1 - [6(\Sigma d^2)] / [n(n^2-1)]$$
$$= 1 - [6(36)] / [8(63)]$$
$$= 1 - .429$$
$$= .571$$

$H_o: \rho_s = 0$
$H_1: \rho_s \neq 0$
$\alpha = .05$
C.R. $r_s < -.738$
$\quad\quad r_s > .738$
calculations:
$\quad r_s = .571$
conclusion:
Do not reject H_o; there is not sufficient evidence to conclude that $\rho_s \neq 0$.

9. The following table summarizes the calculations.

x	R_x	y	R_y	d	d^2
107	10	111	10.5	-0.5	0.25
96	5	97	3	2	4
103	8	116	12	-4	16
90	3	107	8	-5	25
96	5	99	4	1	1
113	12	111	10.5	1.5	2.25
86	1	85	1	0	0
99	7	108	9	-2	4
109	11	102	6	5	25
105	9	105	7	2	4
96	5	100	5	0	0
89	2	93	2	0	0
				0.0	81.50

$$r_s = 1 - [6(\Sigma d^2)]/[n(n^2-1)]$$
$$= 1 - [6(81.5)]/[12(143)]$$
$$= 1 - .285$$
$$= .715$$

$H_o: \rho_s = 0$
$H_1: \rho_s \neq 0$
$\alpha = .05$
C.R. $r_s < -.591$
    $r_s > .591$
calculations:
   $r_s = .715$
conclusion:

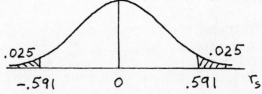

Reject H_o; there is sufficient evidence to conclude that $\rho_s \neq 0$ (in fact, $\rho_s > 0$).

NOTE: Exercise 9 as presented is decidedly different from all the other exercises in this section, and the above test may be open to challenge. When finding the correlation between x and y, it makes no difference which variable carries which designation -- BUT THE PROBLEM MUST BE WELL-DEFINED AND INTERNALLY CONSISTENT. If x is "height" and y is "weight," for example, that designation must be kept for all pairs and one cannot rank one person's height with the other weights. Similarly if x is "test 1" and y is "test 2." Here, however, x and y are assumed to be "twin 1" and "twin 2" -- but how is it determined which twin is which? The problem could be made well-defined and internally consistent by declaring, for example, that x is the "older twin" -- but this was not stated in the exercise. Interchanging which twin is x and which is y for twins 1,2,4,9 produces the results and the same data lead to the opposite conclusion! As in all the other exercises, there should be a clear and consistent determination of what is x and what is y. Even if the choice of x and y was made randomly, that fact should be reported with the data.

x	R_x	y	R_y	d	d^2
111	11	107	8	3	9
97	5	96	4	1	1
103	8	116	12	-4	16
107	10	90	2	8	64
96	3.5	99	5	-1.5	2.25
113	12	111	11	1	1
86	1	85	1	0	0
99	6	108	9	-3	9
102	7	109	10	-3	9
105	9	105	7	2	4
96	3.5	100	6	-2.5	6.25
89	2	93	3	-1	1
				0.0	122.50

$$r_s = 1 - [6(\Sigma d^2)]/[n(n^2-1)]$$
$$= 1 - [6(122.5)]/[12(143)]$$
$$= 1 - .428$$
$$= .572$$

11. The following table summarizes the calculations.

x	R_x	y	R_y	d	d^2
.27	1	2	2.5	-1.5	2.25
1.41	3	3	4.5	-1.5	2.25
2.19	5.5	3	4.5	1	1
2.83	7	6	8	-1	1
2.19	5.5	4	6	-0.5	0.25
1.81	4	2	2.5	1.5	2.25
.85	2	1	1	1	1
3.05	8	5	7	1	1
				0.0	11.00

$$r_s = 1 - [6(\Sigma d^2)]/[n(n^2-1)]$$
$$= 1 - [6(11)]/[8(63)]$$
$$= 1 - .131$$
$$= .869$$

$H_o: \rho_s = 0$
$H_1: \rho_s \neq 0$
$\alpha = .05$
C.R. $r_s < -.738$
    $r_s > .738$
calculations:
    $r_s = .869$
conclusion:
    Reject H_o; there is sufficient evidence to conclude that $\rho_s \neq 0$ (in fact, $\rho_s > 0$).

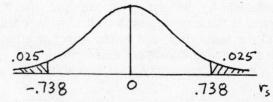

13. The following table summarizes the calculations.

x	R_x	y	R_y	d	d^2
17.2	1	.19	3	-2	4
43.5	6	.20	4.5	1.5	2.25
30.7	4	.26	8	-4	16
53.1	8	.16	1	7	49
37.2	5	.24	7	-2	4
21.0	2	.20	4.5	-2.5	6.25
27.6	3	.18	2	1	1
46.3	7	.23	6	1	1
				0.0	83.50

$$r_s = 1 - [6(\Sigma d^2)]/[n(n^2-1)]$$
$$= 1 - [6(83.5)]/[8(63)]$$
$$= 1 - .994$$
$$= .006$$

$H_o: \rho_s = 0$
$H_1: \rho_s \neq 0$
$\alpha = .05$
C.R. $r_s < -.738$
    $r_s > .738$
calculations:
    $r_s = .006$
conclusion:
    Do not reject H_o; there is not sufficient evidence to conclude that $\rho_s \neq 0$.

15. The following table summarizes the calculations.

x	R_x	y	R_y	d	d^2
17	6	73	8	-2	4
21	8	66	4	4	16
11	2.5	64	3	-0.5	0.25
16	5	61	2	3	9
15	4	70	6	-2	4
11	2.5	71	7	-4.5	20.25
24	9	90	10	-1	1
27	10	68	5	5	25
19	7	84	9	-2	4
8	1	52	1	0	0
				0.0	83.50

$$r_s = 1 - [6(\Sigma d^2)] / [n(n^2-1)]$$
$$= 1 - [6(83.5)] / [10(99)]$$
$$= 1 - .506$$
$$= .494$$

$H_o: \rho_s = 0$
$H_1: \rho_s \neq 0$
$\alpha = .05$
C.R. $r_s < -.648$
$\quad r_s > .648$
calculations:
$\quad r_s = .494$
conclusion:
$\quad$ Do not reject H_o; there is not sufficient evidence to conclude that $\rho_s \neq 0$.

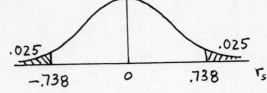

17. The following table summarizes the calculations.

x	R_x	y	R_y	d	d^2
.65	6	14.7	7	-1	1
.55	4	12.3	5	-1	1
.72	7	14.6	6	1	1
.83	8	15.1	8	0	0
.57	5	5.0	4	1	1
.51	3	4.1	2.5	0.5	0.25
.43	2	3.8	1	1	1
.37	1	4.1	2.5	-1.5	2.25
				0.0	7.50

$$r_s = 1 - [6(\Sigma d^2)] / [n(n^2-1)]$$
$$= 1 - [6(7.5)] / [8(63)]$$
$$= 1 - .089$$
$$= .911$$

$H_o: \rho_s = 0$
$H_1: \rho_s \neq 0$
$\alpha = .05$
C.R. $r_s < -.738$
$\quad r_s > .738$
calculations:
$\quad r_s = .911$
conclusion:
$\quad$ Reject H_o; there is sufficient evidence to conclude that $\rho_s \neq 0$ (in fact, $\rho_s > 0$).

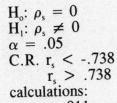

19. The following table summarizes the calculations.

x	R_x	y	R_y	d	d^2
3522	1	.20	1	0	0
3597	2	.22	2	0	0
4171	7	.23	3	4	16
4258	8	.29	4	4	16
3993	4	.31	5	-1	1
3971	3	.33	7.5	-4.5	20.25
4042	5	.33	7.5	-2.5	6.25
4053	6	.32	6	0	0
				0.0	59.50

$$r_s = 1 - [6(\Sigma d^2)]/[n(n^2-1)]$$
$$= 1 - [6(59.5)]/[8(63)]$$
$$= 1 - .708$$
$$= .292$$

H_o: $\rho_s = 0$
H_1: $\rho_s \neq 0$
$\alpha = .05$
C.R. $r_s < -.738$
$\quad r_s > .738$
calculations:
$\quad r_s = .292$
conclusion:
Do not reject H_o; there is not sufficient evidence to conclude that $\rho_s \neq 0$.

21. The following table summarizes the calculations.

x	R_x	y	R_y	d	d^2
65.5	2	12.6	3	-1	1
68.3	9	36.4	11	-2	4
73.1	17	13.6	4	13	169
71.8	15	43.2	13	2	4
67.3	5	32.4	9	-4	16
70.2	14	34.8	10	4	16
68.7	11	54.6	15	-4	16
68.4	10	11.6	2	8	64
64.8	1	157.8	17	-16	256
70.0	12.5	20.2	7	5.5	30.25
68.0	7	19.4	6	1	1
66.8	4	28.4	8	-4	16
70.0	12.5	89.2	16	-3.5	12.25
65.7	3	11.4	1	2	4
72.0	16	18.8	5	11	121
68.1	8	48.8	14	-6	36
67.6	6	40.4	12	-6	36
					802.50

$$r_s = 1 - [6(\Sigma d^2)]/[n(n^2-1)]$$
$$= 1 - [6(802.5)]/[17(288)]$$
$$= 1 - .983$$
$$= .017$$

H_o: $\rho_s = 0$
H_1: $\rho_s \neq 0$
$\alpha = .05$
C.R. $r_s < -.490$
$\quad r_s > .490$
calculations:
$\quad r_s = .017$
conclusion:
Do not reject H_o; there is not sufficient evidence to conclude that $\rho_s \neq 0$.

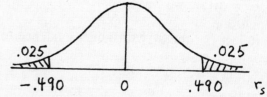

23. The following table summarizes the calculations.

x	R_x	y	R_y	d	d^2
2948	18	189.1	19.5	-1.5	2.25
3536	27	205.2	29	-2	4
3472	26	194.8	24	2	4
2782	12	189.1	19.5	-7.5	56.25
3766	30	178.7	5	25	625
4367	32	225.1	32	0	0
2649	6	183.4	13.5	-7.5	56.25
2526	3	182.3	12	-9	81
2665	7	183.4	13.5	-6.5	42.25
3374	25	199.3	25	0	0
2863	17	184.8	15.5	1.5	2.25
3010	20	184.8	15.5	4.5	20.25
2779	11	179.0	7	4	16
3315	23	201.7	27	-4	16
2788	14	181.2	10.5	3.5	12.25
3290	22	201.6	26	-4	16
2360	2	170.9	2	0	0
2775	10	179.6	8	2	4
2619	5	178.9	6	-1	1
3253	21	193.1	21	0	0
1650	1	147.4	1	0	0
3628	28	205.1	28	0	0
3784	31	212.4	31	0	0
2602	4	177.0	4	0	0
2717	9	187.9	18	-9	81
2992	19	194.4	23	-4	16
3354	24	193.7	22	2	4
3697	29	205.5	30	-1	1
2784	13	181.2	10.5	2.5	6.25
2804	15	186.9	17	-2	4
2823	16	176.3	3	13	169
2682	8	180.7	9	-1	1
				0.0	1241.00

$$r_s = 1 - [6(\Sigma d^2)] / [n(n^2-1)]$$
$$= 1 - [6(1241)] / [32(1023)]$$
$$= 1 - .227$$
$$= .773$$

since n > 30,
$$CV: \pm z_{\alpha/2}/\sqrt{n-1} = \pm 1.960/\sqrt{31}$$
$$= \pm .352$$

$H_o: \rho_s = 0$
$H_1: \rho_s \neq 0$
$\alpha = .05$
C.R. $r_s < -.352$
$\qquad r_s > .352$
calculations:
$\qquad r_s = .773$
conclusion:
    Reject H_o; there is sufficient evidence to conclude that $\rho_s \neq 0$ (in fact, $\rho_s > 0$).

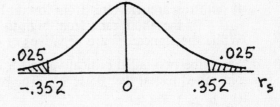

25. a. Excluding ties, the six possible rankings for the second judge are 123 132 213 231 312 321.

   b. The following table organizes the calculations.

2nd judge	d	d^2	Σd^2	$r_s = 1 - [6(\Sigma d^2)]/[n(n^2-1)]$
123	0,0,0	0,0,0	0	$1 - [6 \cdot 0]/[3 \cdot 8] = 1 - 0 = 1$
132	0,-1,1	0,1,1	2	$1 - [6 \cdot 2]/[3 \cdot 8] = 1 - .5 = .5$
213	-1,1,0	1,1,0	2	$1 - [6 \cdot 2]/[3 \cdot 8] = 1 - .5 = .5$
231	-1,-1,2	1,1,4	6	$1 - [6 \cdot 6]/[3 \cdot 8] = 1 - 1.5 = -.5$
312	-2,1,1	4,1,1	6	$1 - [6 \cdot 6]/[3 \cdot 8] = 1 - 1.5 = -.5$
321	-2,0,2	4,0,4	8	$1 - [6 \cdot 8]/[3 \cdot 8] = 1 - 2 = -1$

   c. $P(r_s > .9) = P(r_s = 1) = 1/6 = .167$

27. a. For the pictured data, r_s would give a better measure of the relationship than would r. Since the relationship is non-linear, the linear correlation coefficient r cannot accurately measure its strength. It is true, however, that there is a positive correlation -- i.e., smaller x's are associated with smaller y's and larger x's are associated with larger y's. The rank correlation coefficient r_s measures precisely such an association -- independent of the magnitude of the increases (i.e., whether the increase is linear, quadratic or irregular).

   b. When dealing with n ranks, $\Sigma R = n(n+1)/2$ -- this is always true (i.e., whether or not there are ties in the ranks) and may be used as a check to see if the ranks have been assigned correctly. It is also true that $\Sigma R^2 = n(n+1)(2n+1)/6$ whenever there are no ties in ranks, and that ΣR^2 is less than that whenever there are ties. The shortcut formula
   $$r_s = 1 - [6(\Sigma d^2)]/[n(n^2-1)]$$
   was derived using $\Sigma R^2 = n(n+1)(2n+1)/6$ (i.e., assuming there were no ties). If x is ranked from low to high to produce the usual R_x values and y is ranked from high to low, each y rank will be $(n+1) - R_y$ -- where R_y represents the usual low to high ranking.

   Instead of $d = R_x - R_y$,
   we now have $d = R_x - [(n+1) - R_y]$.
   If there are no ties, algebra and the above formulas for ΣR and ΣR^2 can be used to show the resulting r_s will be precisely the negative of the r_s obtained by ranking both variables from low to high. Whether or not there are ties, the "true" formula always produces that result -

   viz., that ranking one of the variables from high to low reverses the sign of r_s.

   c. Refer to the comments and notation in part (b).
   If both researchers rank from low to high, then $d = R_x - R_y$;
         if they both rank from high to low, then $d = [(n+1) - R_x] - [(n+1) - R_y] = R_y - R_x$.
   While the signed d's are negatives of each other. the d^2 values and r_s remain unchanged.

   d. Since the two-tailed critical value for $\alpha = .10$ places .05 in each tail, it may be used for a one-tailed test at the .05 level. The requested test is as follows.

   $H_o: \rho_s \leq 0$
   $H_1: \rho_s > 0$
   $\alpha = .05$
   C.R. $r_s > .564$
   calculations:
      $r_s = .855$
   conclusion:
      Reject H_o; there is sufficient evidence to conclude that $\rho_s > 0$.

13-7 Runs Test for Randomness

NOTE: In each exercise, the item that appears first in the sequence is considered to be of the first type and its count is designated by n_1.

1. $n_1 = 7$ (# of A's)
   $n_2 = 2$ (# of B's)
   $G = 2$ (# of runs -- one run of A's, followed by one run of B's)
   CV: 1,6 (from Table A-10)

3. a. In increasing order, the values are 3 7 7 8 9 10 12 16 18 20
   and the median is $(9 + 10)/2 = 9.5$

   b. The original sequence 3 8 7 7 9 12 10 16 20 18 becomes the following B's and A's.
      B B B B B A A A A A

   c. $n_1 = 5$
      $n_2 = 5$
      $G = 2$

   d. CV: 2,10

   e. Since $G \leq 2$, we conclude there is not a random sequence of values above and below the median.

5. Since $n_1 = 6$ and $n_2 = 24$, use the normal approximation.
   $$\mu_G = 2n_1 n_2/(n_1 + n_2) + 1$$
   $$= 2(6)(24)/30 + 1$$
   $$= 10.6$$
   $$\sigma_G^2 = [2n_1 n_2(2n_1 n_2 - n_1 - n_2)]/[(n_1 + n_2)^2(n_1 + n_2 - 1)]$$
   $$= [2(6)(24)(258)]/[(30)^2(29)]$$
   $$= 2.847$$

   H_o: the sequence is random
   H_1: the sequence is not random
   $\alpha = .05$
   C.R. $z < -z_{.025} = -1.960$
   $\quad\quad z > z_{.025} = 1.960$
   calculations:

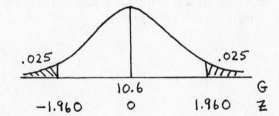

   $\quad G = 12$
   $\quad z_G = (G - \mu_G)/\sigma_G$
   $\quad\quad = (12 - 10.6)/\sqrt{2.847}$
   $\quad\quad = 1.4/1.687$
   $\quad\quad = .830$
   conclusion:
   Do not reject H_o; there is not sufficient evidence to conclude that the sequence is not random.

7. When the scores are arranged in order, the median is found to be $(.14 + .15)/2 = .145$.
In the usual above-below notation, the original sequence may be written as
A A A A A A A A A B A B B B B B B B B B
Since $n_1 = 10$ and $n_2 = 10$, use Table A-10.

H_o: values above and below the median occur in a random sequence
H_1: values above and below the median do not occur in a random sequence
$\alpha = .05$
C.R. $G \leq 6$
     $G \geq 16$
calculations:
    $G = 4$

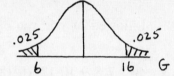

conclusion:
Reject H_o; there is sufficient evidence to conclude that the values above and below the median do not occur in a random sequence (in fact, values on the same side of the median tend to occur in groups).

9. When the scores are arranged in order, the median is found to be $(812 + 856)/2 = 834$.
In the usual above-below notation, the original sequence may be written as
A A A A A A A A B B B B B B B B
Since $n_1 = 8$ and $n_2 = 8$, use Table A-10.

H_o: values above and below the median occur in a random sequence
H_1: values above and below the median do not occur in a random sequence
$\alpha = .05$
C.R. $G \leq 4$
     $G \geq 14$
calculations:
    $G = 2$

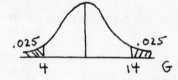

conclusion:
Reject H_o; there is sufficient evidence to conclude that the values above and below the median do not occur in a random sequence (in fact, values on the same side of the median tend to occur in groups).

11. The sequence in Y's and O's is as follows.

    Y Y Y Y Y Y O Y O O O Y Y Y Y O
    O O O O O O O O O O O O O O Y Y
    Y O O O O Y Y O Y O O O O O O O

Since $n_1 = 17$ and $n_2 = 28$, use the normal approximation.

$\mu_G = 2n_1n_2/(n_1+n_2) + 1$
$= 2(17)(28)/45 + 1$
$= 22.156$
$\sigma_G^2 = [2n_1n_2(2n_1n_2-n_1-n_2)]/[(n_1+n_2)^2(n_1+n_2-1)]$
$= [2(17)(28)(907)]/[(45)^2(44)]$
$= 9.691$

H_o: the sequence is random
H_1: the sequence is not random
$\alpha = .05$
C.R. $z < -z_{.025} = -1.960$
    $z > z_{.025} = 1.960$
calculations:
    $G = 12$
    $z_G = (G - \mu_G)/\sigma_G$
    $= (12 - 22.156)/\sqrt{9.691}$
    $= -10.156/3.113$
    $= -3.262$

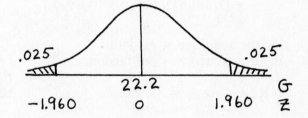

conclusion:
    Reject H_o; there is sufficient evidence to conclude that the sequence is not random (in fact, cases from the same age level tend to occur in groups).

13. Since $n_1 = 50$ and $n_2 = 37$, use the normal approximation.

$\mu_G = 2n_1n_2/(n_1+n_2) + 1$
$= 2(50)(37)/87 + 1$
$= 43.529$
$\sigma_G^2 = [2n_1n_2(2n_1n_2-n_1-n_2)]/[(n_1+n_2)^2(n_1+n_2-1)]$
$= [2(50)(37)(3613)]/[(87)^2(86)]$
$= 20.537$

H_o: the sequence is random
H_1: the sequence is not random
$\alpha = .05$
C.R. $z < -z_{.025} = -1.960$
    $z > z_{.025} = 1.960$
calculations:
    $G = 48$
    $z_G = (G - \mu_G)/\sigma_G$
    $= (48 - 43.529)/\sqrt{20.54}$
    $= 4.471/4.532$
    $= .987$

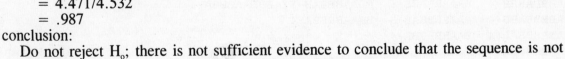

conclusion:
    Do not reject H_o; there is not sufficient evidence to conclude that the sequence is not random.

15. The sequence in O's and E's is as follows.

O E O O O E E O O O O E O O O O O E O E E E
E E E O O E O E O E O O O E E E E E O O O O
E O O O O O O O O E O E E E E O O E O E O E E
O O E O E O E O E O E E E E E E E E E E O O
E E E E E O E E E O O E E O O O E E O O

Since $n_1 = 49$ and $n_2 = 51$, use the normal approximation.

$\mu_G = 2n_1 n_2/(n_1 + n_2) + 1$
$\quad = 2(49)(51)/100 + 1$
$\quad = 50.98$

$\sigma_G^2 = [2n_1 n_2 (2n_1 n_2 - n_1 - n_2)]/[(n_1 + n_2)^2 (n_1 + n_2 - 1)]$
$\quad = [2(49)(51)(4898)]/[(100)^2(99)]$
$\quad = 24.727$

H_o: the sequence is random
H_1: the sequence is not random
$\alpha = .05$
C.R. $z < -z_{.025} = -1.960$
$\quad\quad z > z_{.025} = 1.960$
calculations:
$\quad G = 43$
$\quad z_G = (G - \mu_G)/\sigma_G$
$\quad\quad = (43 - 50.98)/\sqrt{24.73}$
$\quad\quad = -7.98/4.973$
$\quad\quad = -1.605$

conclusion:
Do not reject H_o; there is not sufficient evidence to conclude that the sequence is not random.

17. The minimum possible number of runs is $G = 2$ and occurs when all the A's are together and all the B's are together (e.g., A A B B).

The maximum possible number of runs is $G = 4$ and occurs when the A's and B's alternate (e.g., A B A B).

Because the critical region for $n_1 = n_2 = 2$ is

$\quad$ C.R. $G \leq 1$
$\quad\quad\quad G \geq 6$

the null hypothesis of the sequence being random can never be rejected at the .05 level. Very simply, this means that it is not possible for such a small sample to provide 95% certainty that a non-random phenomenon is occurring.

19. a. and b. The 84 sequences and the number of runs in each are as follows.

AAABBBBB-2	BAAABBBBB-3	BBAAABBBB-3	BBBAAABBB-3	BBBBAAABB-3	BBBBBAAAB-3
AABABBBBB-4	BAABABBBB-5	BBAABABBB-5	BBBAABABB-5	BBBBAABAB-5	BBBBBAABA-4
AABBABBBB-4	BAABBABBB-5	BBAABBABB-5	BBBAABBAB-5	BBBBAABBA-4	
AABBBABBB-4	BAABBBABB-5	BBAABBBAB-5	BBBAABBBA-4		
AABBBBABB-4	BAABBBBAB-5	BBAABBBBA-4			
AABBBBBAB-4	BAABBBBBA-4				
AABBBBBBA-3					BBBBBBAAA-2

```
ABAABBBBB-4    BABAABBBB-5    BBABAABBB-5    BBBABAABB-5    BBBBABAAB-5    BBBBBABAA-4
ABABABBBB-6    BABABABBB-7    BBABABABB-7    BBBABABAB-7    BBBBABABA-6
ABABBABBB-6    BABABBABB-7    BBABABBAB-7    BBBABABBA-6
ABABBBABB-6    BABABBBAB-7    BBABABBBA-6
ABABBBBAB-6    BABABBBBA-6
ABABBBBBA-5
ABBAABBBB-4    BABBAABBB-5    BBABBAABB-5    BBBABBAAB-5    BBBBBABBAA-4
ABBABABBB-6    BABBABABB-7    BBABBABAB-7    BBBABBABA-6
ABBABBABB-6    BABBABBAB-7    BBABBABBA-6
ABBABBBAB-6    BABBABBBA-6
ABBABBBBA-5
ABBBAABBB-4    BABBBAABB-5    BBABBBAAB-5    BBBABBBAA-4
ABBBABABB-6    BABBBABAB-7    BBABBBABA-6
ABBBABBAB-6    BABBBABBA-6
ABBBABBBA-5
ABBBBAABB-4    BABBBBAAB-5    BBABBBBAA-4
ABBBBABAB-6    BABBBBABA-6
ABBBBABBA-4
ABBBBBAAB-4    BABBBBBAA-4
ABBBBBABA-5
ABBBBBBAA-3
```

c. Below is the distribution for G, the number of runs, found from the above sequences.

G	P(G)	
2	2/84 =	.023
3	7/84 =	.083
4	21/84 =	.250
5	24/84 =	.286
6	20/84 =	.238
7	10/84 =	.229
	84/84 =	1.000

d. Based on the above distribution, a two-tailed test at the .05 level (that places .025 or less in each tail) has

    C.R. $G \leq 2$

          $G \geq 8$

(for which it will never be possible to reject by being in the upper tail).

e. The critical region in part (d) agrees exactly with Table A-10.

f. From the distribution in part (c)...

G	P(G)	$G \cdot P(G)$
2	2/84	4/84
3	7/84	21/84
4	21/84	84/84
5	24/84	120/84
6	20/84	120/84
7	10/84	70/84
		419/84

$$\mu_G = \Sigma G \cdot P(G)$$
$$= 419/84$$

From Formula 13-2...

$$\mu_G = 2n_1 n_2 / (n_1 + n_2) + 1$$
$$= 2(3)(6)/(9) + 1$$
$$= 5$$
$$[= 420/84]$$

NOTE: In most large-sample formulas in this chapter, the given μ and σ are exact and the distribution is approximately normal. In this section, the given μ and σ are approximations that improve as n_1 and n_2 increase -- and even for the small samples here the error is only 1/84.

Review Exercises

1. Since $n_1 = 20$ and $n_2 = 5$, use Table A-10.

   H_o: the sequence is random
   H_1: the sequence is not random
   $\alpha = .05$ [only value in Table A-10]
   C.R. $G \leq 5$
   $\quad\quad G \geq 12$
   calculations:
   $\quad G = 5$
   conclusion:

   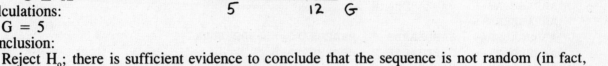

   Reject H_o; there is sufficient evidence to conclude that the sequence is not random (in fact, similar types tend to be grouped together).

3. The following table summarizes the calculations.

R_x	R_y	d	d^2
1	4	-3	9
4	2	2	4
3	1	2	4
5	6	-1	1
6	3	3	9
2	7	-5	25
7	5	2	4
		0	56

$$r_s = 1 - [6(\Sigma d^2)]/[n(n^2-1)]$$
$$= 1 - [6(56)]/[7(48)]$$
$$= 1 - 1$$
$$= 0$$

H_o: $\rho_s = 0$
H_1: $\rho_s \neq 0$
$\alpha = .05$
C.R. $r_s < -.786$
$\quad\quad r_s > .786$
calculations:
$\quad r_s = 0$
conclusion:

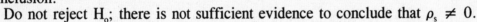

Do not reject H_o; there is not sufficient evidence to conclude that $\rho_s \neq 0$.

5. Below are the scores (in order) for each group.

1	R	2	R	3	R
160	1	165	3	162	2
172	4	174	6	175	7
173	5	176	9	177	11.5
176	9	180	16.5	179	14.5
176	9	181	18	187	23
177	11.5	184	20	195	27.5
178	13	186	22	210	31
179	14.5	190	24	215	32
180	16.5	192	26	216	33
183	19	195	27.5	220	34
185	21	200	29	222	35
191	25	201	30		250.5
	148.5		231.0		

$n_1 = 12$ $R_1 = 148.5$
$n_2 = 12$ $R_2 = 231.0$
$n_3 = 11$ $R_3 = 250.5$

$n = \Sigma n = 35$ $\Sigma R = 630.0$

check:
$\Sigma R = n(n+1)/2$
$= 35(36)/2$
$= 630$

H_o: the populations have the same distribution
H_1: the populations have different distributions
$\alpha = .05$
C.R. $H > \chi^2_{2,.05} = 5.991$

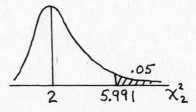

calculations:

$H = [12/n(n+1)] \cdot [\Sigma(R_i^2/n_i)] - 3(n+1)$
$= [12/35(36)] \cdot [(148.5)^2/12 + (231)^2/12 + (250.5)^2/11] - 3(36)$
$= [.00952] \cdot [11989] - 108$
$= 6.181$

conclusion:
Reject H_o; there is sufficient evidence to conclude that the populations have different distributions.

7. Let Test A be group 1.
   claim: median difference = 0

pair	1	2	3	4	5	6	7	8	9
H-S	-	-	-	+	-	-	-	+	-

n = 9 +'s and -'s

H_o: median difference = 0
H_1: median difference $\neq$ 0
$\alpha = .05$

C.R. $x \le x_{L,9,.025} = 1$ OR   C.R. $x \le x_{L,9,.025} = 1$
$x \ge x_{U,9,.025} = 9-1 = 8$

calculations: calculations:
  x = 2   x = 2

conclusion:
Do not reject H_o; there is not sufficient evidence to conclude that median difference $\neq$ 0.

9. Below are the scores (in order) for each group. The group listed first is considered group 1.

Ora	R	Wes	R
14	4	5	1
26	6	7	2
39	9.5	10	3
60	16	17	5
62	17	27	7
63	18	35	8
66	19	39	9.5
70	20	40	11
75	21	48	12
79	23	49	13
86	24	50	14
	177.5	54	15
		78	22
			122.5

$n_1 = 11$ $\qquad \Sigma R_1 = 177.5$
$n_2 = 13$ $\qquad \Sigma R_2 = 122.5$

$n = \Sigma n = 24$ $\qquad \Sigma R = 300.0$

check: $\Sigma R = n(n+1)/2$
$\qquad\qquad = 24(25)/2$
$\qquad\qquad = 300$

$R = \Sigma R_1 = 177.5$

$\mu_R = n_1(n+1)/2$
$\qquad = 11(25)/2$
$\qquad = 137.5$

$\sigma_R^2 = n_1 n_2 (n+1)/12$
$\qquad = (11)(13)(25)/12$
$\qquad = 297.917$

H_o: the populations have the same distribution
H_1: the populations have different distributions
$\alpha = .05$
C.R. $z < -z_{.025} = -1.960$
$\qquad z > z_{.025} = 1.960$
calculations:
$\qquad z_R = (R - \mu_R)/\sigma_R$
$\qquad\quad = (177.5 - 137.5)/\sqrt{297.917}$
$\qquad\quad = 40/17.260$
$\qquad\quad = 2.317$

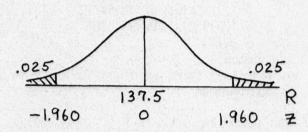

conclusion:
$\quad$ Reject H_o; there is sufficient evidence to conclude that the populations have different distributions (in fact, the Orange County scores tend to be higher).

11. The following table summarizes the calculations.

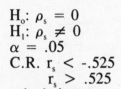

R_x	R_y	d	d^2
1	1	0	0
3	2	1	1
2	3	-1	1
7	8	-1	1
6	6.	0	0
10	7	3	9
14	14	0	0
15	15	0	0
11	4	7	49
9	10	-1	1
12	13	-1	1
4	5	-1	1
8	9	-1	1
5	11	-6	36
13	12	1	1
		0	102

$$r_s = 1 - [6(\Sigma d^2)]/[n(n^2-1)]$$
$$= 1 - [6(102)]/[15(224)]$$
$$= 1 - .182$$
$$= .818$$

H_o: $\rho_s = 0$
H_1: $\rho_s \neq 0$
$\alpha = .05$
C.R. $r_s < -.525$
$\quad\quad r_s > .525$
calculations:
$\quad r_s = .818$
conclusion:
$\quad$ Reject H_o; there is sufficient evidence to conclude that $\rho_s \neq 0$ (in fact, $\rho_s > 0$).

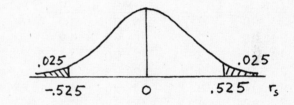

13. Let Before be group 1.
    claim: the populations have the same distribution

pair	1	2	3	4	5	6	7	8	9	10	11	12	13	14
I-II	-2	0	-4	-4	-1	2	-4	-3	0	-1	-5	-2	-5	-3
R	-4	0	-9	-9	-1.5	4	-9	-6.5	0	-1.5	-11.5	-4	-11.5	-6.5

n = 12 non-zero ranks
$\Sigma R- = 74.0$
$\Sigma R+ = 4.0$
$\Sigma R = 78.0$ [check: $\Sigma R = n(n+1)/2 = 12(13)/2 = 78$]

H_o: the populations have the same distribution
H_1: the populations have different distributions
$\alpha = .05$
C.R. $T \leq T_{L,12,.025} = 14$ <u>OR</u>   C.R. $T \leq T_{L,12,.025} = 14$
$\quad\quad\quad\quad\quad\quad\quad\quad\quad\quad\quad\quad\quad\quad\quad\quad T \geq T_{U,12,.025} = 78\text{-}14 = 64$

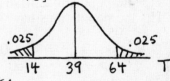

calculations: calculations:
$\quad T = 4$ $\quad T = 4$
conclusion:
$\quad$ Reject H_o; there is sufficient evidence to conclude that the populations have different distributions (in fact, the Before scores tend to be lower).

15. Let Before be group 1.
claim: median difference = 0

pair	1	2	3	4	5	6	7	8	9	10	11	12	13	14
H-S	-	0	-	-	-	+	-	-	0	-	-	-	-	-

n = 12 +'s and -'s

H_o: median difference = 0
H_1: median difference $\neq$ 0
α = .05

C.R. $x \leq x_{L,12,.025} = 2$ __OR__   C.R. $x \leq x_{L,12,.025} = 2$
                                                    $x \geq x_{U,12,.025} = 12\text{-}2 = 10$

calculations: calculations:
   x = 1   x = 1

conclusion:
Reject H_o; there is sufficient evidence to conclude that median difference $\neq$ 0 (in fact, median difference < 0 -- i.e., the Before scores tend to be lower).

17. Below are the scores (in order) for each group. The group listed first is considered group 1.

MO	R	CS	R
23.00	1	29.00	10
24.50	2	30.99	12
24.75	3	32.00	14
26.00	4	32.99	16
27.00	5	33.00	17
27.98	6	33.98	18
27.99	7	33.99	19
28.15	8	34.79	20
29.99	10	35.79	21
29.99	10	37.75	22
31.50	13	38.99	30
32.75	15		192
	84		

$n_1 = 12$ $\Sigma R_1 = 84$
$n_2 = 11$ $\Sigma R_2 = 192$

n = Σn = 23 ΣR = 276

check: $\Sigma R = n(n+1)/2$
            = 23(24)/2
            = 276

$R = \Sigma R_1 = 84$

$\mu_R = n_1(n+1)/2$
     = 12(24)/2
     = 144

$\sigma_R^2 = n_1 n_2 (n+1)/12$
     = (11)(12)(24)/12
     = 264

H_o: the populations have the same distribution
H_1: the populations have different distributions
α = .05
C.R. $z < -z_{.025} = -1.960$
      $z > z_{.025} = 1.960$
calculations:
   $z_R = (R - \mu_R)/\sigma_R$
      $= (84 - 144)/\sqrt{264}$
      = -60/16.248
      = -3.693

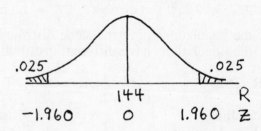

conclusion:
Reject H_o; there is sufficient evidence to conclude that the populations have different distributions (in fact, the mail order prices are lower).

19. Below are the scores (in order) for each group.

1	R	2	R	3	R
195	3	187	1	193	2
198	4	210	5	212	6
223	9	222	8	215	7
240	12	238	11	231	10
251	13	256	15	252	14
	41		40	260	16
				267	17
					72

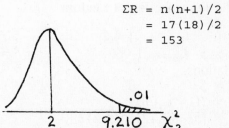

$n_1 = 5$ $R_1 = 41$
$n_2 = 5$ $R_2 = 40$
$n_3 = 7$ $R_3 = 72$

$n = \Sigma n = 17$ $\Sigma R = 153$

check:
$\Sigma R = n(n+1)/2$
$= 17(18)/2$
$= 153$

H_o: the populations have the same distribution
H_1: the populations have different distributions
$\alpha = .05$
C.R. $H > \chi^2_{2,.01} = 9.210$
calculations:

$H = [12/n(n+1)] \cdot [\Sigma(R_i^2/n_i)] - 3(n+1)$
$= [12/17(18)] \cdot [(41)^2/5 + (40)^2/5 + (72)^2/7] - 3(18)$
$= [.0392] \cdot [1396.77] - 54$
$= .775$

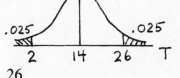

conclusion:
Do not reject H_o; there is not sufficient evidence to conclude that the populations have different distributions.

21. Let Judge A be group 1.
claim: the populations have the same distribution

pair	1	2	3	4	5	6	7	8
I-II	-.8	.7	-1.6	-.1	.7	.9	1.0	0
R	-4	2.5	-7	-1	2.5	5	6	0

$n = 7$ non-zero ranks
$\Sigma R- = 12.0$
$\Sigma R+ = 16.0$
$\Sigma R = 28.0$ [check: $\Sigma R = n(n+1)/2 = 7(8)/2 = 28$]

H_o: the populations have the same distribution
H_1: the populations have different distributions
$\alpha = .05$
C.R. $T \leq T_{L,7,.025} = 2$ OR C.R. $T \leq T_{L,7,.025} = 2$
$T \geq T_{U,7,.025} = 28-2 = 26$

calculations: calculations:
$T = 12$ $T = 16$

conclusion:
Do not reject H_o; there is not sufficient evidence to conclude that the populations have different distributions.

NOTE: This test may not be the most appropriate one for the intended purpose. The signed-rank test measures overall differences between the judges and not whether the two judges give the same results for individuals. If one judge gave half the students higher scores and the other half of the students scores that were lower by the same amounts, then $\Sigma R-$ would equal $\Sigma R+$ (so we could not reject H_o) but the distributions would be very different. Thus the test measures whether the overall typical scores are the same for the two judges and not whether the two judges agreed on a contestant-by-contestant basis.

23. Since $n_1 = 25$ and $n_2 = 8$, use the normal approximation.

$$\mu_G = 2n_1n_2/(n_1+n_2) + 1$$
$$= 2(25)(8)/33 + 1$$
$$= 13.121$$
$$\sigma_G^2 = [2n_1n_2(2n_1n_2-n_1-n_2)]/[(n_1+n_2)^2(n_1+n_2-1)]$$
$$= [2(25)(8)(367)]/[(33)^2(32)]$$
$$= 4.212$$

H_o: the sequence is random
H_1: the sequence is not random
$\alpha = .05$
C.R. $z < -z_{.025} = -1.960$
$\quad\quad z > z_{.025} = 1.960$

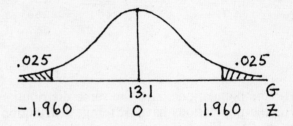

calculations:
$\quad G = 5$
$\quad z_G = (G - \mu_G)/\sigma_G$
$\quad\quad = (5 - 13.121)/\sqrt{4.212}$
$\quad\quad = -8.121/2.052$
$\quad\quad = -3.957$

conclusion:
Reject H_o; there is sufficient evidence to conclude that the sequence is not random (in fact, the N and Y responses tend to occur in groups).